Advances in Industrial Control

T0191590

Springer

London
Berlin
Heidelberg
New York
Barcelona
Hong Kong
Milan
Paris
Singapore
Tokyo

Other titles published in this Series:

Compressor Surge and Rotating Stall: Modeling and Control
Jan Tommy Gravdahl and Olav Egeland

Radiotherapy Treatment Planning: New System Approaches
Olivier Haas

Feedback Control Theory for Dynamic Traffic Assignment
Pushkin Kachroo and Kaan Özbay

Autotuning of PID Controllers
Cheng-Ching Yu

Robust Aeroservoelastic Stability Analysis
Rick Lind and Marty Brenner

Performance Assessment of Control Loops: Theory and Applications
Biao Huang and Sirish L. Shah

Data Mining and Knowledge Discovery for Process Monitoring and Control
Xue Z. Wang

Advances in PID Control
Tan Kok Kiong, Wang Quing-Guo and Hang Chang Chieh with Tore J. Hägglund

Advanced Control with Recurrent High-order Neural Networks: Theory and Industrial Applications
George A. Rovithakis and Manolis A. Christodoulou

Structure and Synthesis of PID Controllers
Aniruddha Datta, Ming-Tzu Ho and Shankar P. Bhattacharyya

Data-driven Techniques for Fault Detection and Diagnosis in Chemical Processes
Evan L. Russell, Leo H. Chiang and Richard D. Braatz

Bounded Dynamic Stochastic Systems: Modelling and Control
Hong Wang

Non-linear Model-based Process Control
Rashid M. Ansari and Moses O. Tadé

Identification and Control of Sheet and Film Processes
Andrew P. Featherstone, Jeremy G. VanAntwerp and Richard D. Braatz

Precision Motion Control: Design and Implementation
Tan Kok Kiong, Lee Tong Heng, Dou Huifang and Huang Sunan

Nonlinear Identification and Control: A Neural Network Approach
Guoping Liu

Digital Controller Implementation and Fragility: A Modern Perspective
Robert S.H. Istepanian and James F. Whidborne

Optimisation of Industrial Processes at Supervisory Level
Doris Sáez, Aldo Cipriano and Andrzej W. Ordys

Huang Sunan, Tan Kok Kiong
and Lee Tong Heng

Applied Predictive Control

With 87 Figures

 Springer

Huang Sunan, PhD
Tan Kok Kiong, PhD
Lee Tong Heng, PhD
Department of Electrical & Computer Engineering,
National University of Singapore, 4 Engineering Drive 3, Singapore 117576

ISBN 978-1-84996-864-5

British Library Cataloguing in Publication Data
Sunan, Huang
    Applied predictive control. - (Advances in industrial
    control)
    1.Predicitve control
    I.Title II.Kiong, Tan Kok III.Heng, Lee Tong
    629.8

Library of Congress Cataloging-in-Publication Data
Huang, Sunan, 1962-
    Applied predictive control / Huang Sunan, Tan Kok Kiong, and Lee Tong Heng.
        p. cm. -- (Advances in industrial control)
    Includes bibliographical references and index.

    1. Predictive control. 2. Control theory. I. Tan, Kok Kiong, 1967- II. Lee, Tong Heng,
    1958- III. Title. IV. Series.
    TJ217.6 .H83 2001
    629.8--dc21   2001049249

Springer-Verlag London Berlin Heidelberg
a member of BertelsmannSpringer Science+Business Media GmbH
http://www.springer.co.uk

© Springer-Verlag London Limited 2010
Printed in Great Britain

Advances in Industrial Control

Professor Dr -Ing M. Thoma
Institut für Regelungstechnik
Universität Hannover
Appelstr. 11
30167 Hannover
Germany

Professor H. Kimura
Department of Mathematical Engineering and Information Physics
Faculty of Engineering
The University of Tokyo
7-3-1 Hongo
Bunkyo Ku
Tokyo 113
Japan

Professor A.J. Laub
College of Engineering – Dean's Office
University of California
One Shields Avenue
Davis
California 95616-5294
United States of America

Professor J.B. Moore
Department of Systems Engineering
The Australian National University
Research School of Physical Sciences
GPO Box 4
Canberra
ACT 2601
Australia

Dr M.K. Masten
Texas Instruments
2309 Northcrest
Plano
TX 75075
United States of America

Professor Ton Backx
AspenTech Europe B.V.
De Waal 32
NL-5684 PH Best
The Netherlands

SERIES EDITORS' FOREWORD

The series *Advances in Industrial Control* aims to report and encourage technology transfer in control engineering. The rapid development of control technology has an impact on all areas of the control discipline. New theory, new controllers, actuators, sensors, new industrial processes, computer methods, new applications, new philosophies..., new challenges. Much of this development work resides in industrial reports, feasibility study papers and the reports of advanced collaborative projects. The series offers an opportunity for researchers to present an extended exposition of such new work in all aspects of industrial control for wider and rapid dissemination.

The *Advances in Industrial Control* series promotes control techniques, which are used by industry. The series has useful volumes in various aspects of proportional-integral-derivative (PID) control because of the widespread use of PID in applications. Predictive control is another technique that quickly became essential in some sectors of the petro-chemical, and process control industries. It was the ability of the method to incorporate operational constraints that lead to this take-up by industry. The wider industrial applications of predictive control has been slower to develop; indeed some practitioners might argue that this technology transfer step is still active or had only just begun in some industrial sectors.

The *Advances in Industrial Control* series published its first volume on the predictive control paradigm in 1995. This was the volume, *Model Predictive Control in the Process Industries* by E.F. Camacho and C. Bordons. This was a very successful volume, which was transferred to the *Advanced Textbooks in Control and Signal Processing* series after revision. This volume also indicated considerable academic and industrial interest in predictive control methods.

Although predictive control has often appeared as secondary theme in many *Advances in Industrial Control* volumes since, this new book *Applied Predictive Control* by Sunan Huang, Kok Kiong Tan and Tong Heng Lee is the first major entry to the Series devoted to predictive control for many years. The new book is carefully constructed, beginning from dead-time compensation, makes an innovative contribution to predictive PID control and concludes with industrial and experimental case studies. We hope this combination of topics will appeal to industrial and academic control engineers alike.

M.J. Grimble and M.A. Johnson
Industrial Control Centre
Glasgow, Scotland, U.K.

PREFACE

Predictive control is now widely regarded as one of the standard control paradigms for industrial processes. Predictive control has many attractive features. Among them, it is easy to accommodate conflicting regulatory and economic requirements in an optimisation criterion. These features are useful and important to meet the ever-increasing demands in applications. In recent times, variants of predictive control have evolved, which combine with other algorithms to achieve better performance.

This book will serve to cover a comprehensive treatment of the subject matter, including both the fundamentals and state-of-the-art developments in the field of predictive control. It first looks at dead time compensators as the earlier applications of predictive control. Two dead time compensators, namely the Smith-predictor controller (SPC) and the finite spectrum assignment (FSA) controller, are presented. Next, the book progresses to its focus topic on generalised predictive control (GPC). Both the basic and advanced forms of GPC are illustrated, with a strong emphasis towards sound stability and robustness analysis. Finally, a substantial part of the book addresses the application issues of predictive control, providing several interesting case studies for the more application-discerning readers, especially those keenly interested in non-linear control systems. Thus, while the book is written with a view to serve as an advanced control reference on predictive control to researchers, postgraduates and senior undergraduates, it should be equally useful to practitioners who are keen to explore the use of predictive control in real problems. The prerequisites to efficient reading of the book is basic knowledge of a first undergraduate control engineering course.

In what follows, the contents of the book will be briefly reviewed.

Dead Time Compensator

The SPC and FSA control methods are first covered. The basics of SPC are elaborated, highlighting the constraints and complementing this with stability and robustness analysis. Stability analysis forms a substantial section, since it is a major fundamental factor behind the success of any model-based control scheme. The FSA control is presented as an alternative to dead time compensation which does not have some of the constraints of SPC, such as applications to unstable systems. Both the basic form of FSA and a modified form to enable tracking and regulation are presented. Stability analysis with respect to practical stability, robust stability and robust performance are accordingly given for the FSA method.

Generalised Predictive Control

Initial discussions on the fundamentals of GPC will set the foundation for the subsequent design and analysis of the control scheme. Issues of common concern, such as stability, input constraints, and availability of the system states (measured or observed) are addressed systematically. Since most practical control systems are subject to disturbances, a learning algorithm is incorporated into GPC to address this problem. Another prominent part of this topic is devoted to illustrating the equivalence of the GPC approach with an adequate system model to an optimally tuned gain-scheduled proportional-integral-derivative (PID) control system. Incorporating an iterative learning scheme, GPC is also applied to develop three predictive learning control schemes which are new feedforward learning algorithms. Finally, considering an uncertain and/or changing model, an adaptive GPC is presented for a class of non-linear systems, where the estimation convergence and the stability of the closed-loop system are analysed systematically. In most of the control schemes presented, an accompanying section on stability and robustness analysis is usually provided.

Applications

Strong properties achieved via the design methods presented in this book have the potential to expand the application range of feedback controllers to systems where linear or other controllers are inadequate, or are unable to give the desired level of performance. Application examples are given in the book to illustrate the applicability of the various control schemes presented earlier to various systems, including the DC motor, intelligent vehicle control system, injection moulding machine, permanent magnet linear motors, and vacuum distillation column.

This book would not have been possible without the generous assistance of the following colleagues and friends: A/P Wang Qing-Guo, Mr Jiang Xi, Mr Leu Fook Meng, Ms Raihana Ferdous. The authors would like to express their sincere appreciation of their kind assistance provided in the writing of the book. Special thanks to Ms Wong Voon Chin. The authors would also like to thank the National University of Singapore (NUS) for generously funding the project. The editing process would not have been as smooth without the usual generous assistance of Mr Oliver Jackson, Mr Michael Saunders and

Ms Drury Catherine. Finally, we would also like to dedicate the book to our families for their love and support.

CONTENTS

1. Introduction 1

2. Dead Time Compensators 9

   2.1   Smith-predictor Controller 9

      2.1.1   The Basic Smith Control System 10

      2.1.2   Constraints of the SPC............................ 11

      2.1.3   Robust Stability.................................. 11

      2.1.4   Robust Performance 15

   2.2   Finite Spectrum Assignment Control...................... 16

      2.2.1   Finite Spectrum Assignment (FSA) Algorithms 17

      2.2.2   Tracking and Regulation 21

      2.2.3   Practical Stability 27

      2.2.4   Robust Stability.................................. 32

      2.2.5   Performance Robustness 33

3. Generalised Predictive Controller...................... 37

   3.1   Prediction, Control and Stability 37

      3.1.1   System Model and Prediction 37

      3.1.2   Minimisation of the Performance Criterion.......... 39

      3.1.3   Stability Analysis 42

3.2 Predictive Control with Input Constraints 46

   3.2.1 Stability Analysis 47

3.3 Predictive Control with Disturbance Learning 53

   3.3.1 GPC Design with Disturbance 54

   3.3.2 Disturbance Learning Scheme 57

   3.3.3 Simulation Example 61

3.4 Non-linear Predictive Control 62

   3.4.1 System Model 65

   3.4.2 Dynamic Linearisation 68

   3.4.3 Predictive Control Scheme........................ 70

4. Observer-augmented Generalised Predictive Controller ... 73

4.1 Predictive Control Solution for Time-delay Systems 73

   4.1.1 System Model with Time-delay and Non-measurable Noise 74

   4.1.2 Prediction with Input Delay and Noise 74

   4.1.3 Kalman Filter as an Observer 77

   4.1.4 Performance Criterion 78

4.2 Stability and Robustness Analysis 79

   4.2.1 Stability 80

   4.2.2 Robustness 85

4.3 Simulation Study 88

5. Optimal PID Control Based on a Predictive Control Approach 99

5.1 Structure of the PID Controller 100

5.2 Optimal PI Controller Design Based on GPC 101

   5.2.1 GPC Solution for Time-delay Systems 102

5.2.2   Case of a Single-step Control Horizon 105

5.2.3   Stability 105

5.2.4   PI Tuning for First-order Systems 106

5.2.5   Controller Design Based on GPC Approach 106

5.2.6   Selection of Weighting Matrices 111

5.2.7   Stability 113

5.2.8   Robustness Analysis 124

5.2.9   On-line Tuning of the Control Weightings 128

5.2.10 Simulation Examples 129

5.2.11 Real-time Experiments 134

5.3   Optimal PID Controller Design Based on GPC 135

5.3.1   General System Model 136

5.3.2   GPC Control Design 138

5.3.3   Selection of Control Horizon 141

5.3.4   Selection of Weighting Matrices 142

5.3.5   Choice of ζ, ω_n and P_1 143

5.3.6   Q–D Relationship 143

5.3.7   Simulation....................................... 148

6. **Predictive Iterative Learning Control** 153

6.1   Linear Iterative Learning Control Algorithm 154

6.2   Predictive Iterative Learning Control Using Package Information.. 154

6.2.1   System Model and Problem Statements 155

6.2.2   Predictor Construction 156

6.2.3   Derivation of Algorithm 157

6.2.4   Convergence and Robustness of Algorithm 160

6.2.5   Simulation.. 165

6.3   Predictive Iterative Learning Control without Package Infor-
        mation 166

        6.3.1   Problem Statement 168

        6.3.2   Error Propagation 169

        6.3.3   Predictor Construction 170

        6.3.4   Derivation of Predictive Learning Algorithm 172

        6.3.5   Convergence of Predictive Learning Algorithm 172

        6.3.6   Simulation.. 174

6.4   Non-linear Predictive Iterative Learning Control............ 175

        6.4.1   System Model 177

        6.4.2   Dynamic Linearisation 177

        6.4.3   Predictive Learning Control Scheme 178

7.   Adaptive Predictive Control of a Class of SISO Non-linear
     Systems 181

7.1   Design of Adaptive Predictive Controller 181

        7.1.1   System Model 183

        7.1.2   Parameter Estimation............................. 187

        7.1.3   Prediction and Control Action 188

7.2   Stability Analysis 191

        7.2.1   Parameter Identification Convergence 191

        7.2.2   Stability of Adaptive Controller 193

7.3   Simulation... 202

        7.3.1   Linear System 202

        7.3.2   The Injection Moulding Control Problem 202

        7.3.3   Linear Motor Control Problem 204

8. Case Studies 207

   8.1 Introduction 207

   8.2 Application to Intelligent Autonomous Vehicle 208

      8.2.1 Description of Intelligent Vehicle Control 209

      8.2.2 Decision-making 211

      8.2.3 Interface between Decision Level and Control Level ... 218

      8.2.4 Optimal Tracking Based on Predictive Control 220

      8.2.5 Simulations on SmartPath Simulator 230

   8.3 Application to Injection Moulding Control 236

      8.3.1 The Control Problem 237

      8.3.2 Process Model 237

      8.3.3 Controller Design and Experimential Results 242

   8.4 Application to a DC Motor 243

References 249

Index 263

CHAPTER 1

INTRODUCTION

Predictive control, as the name implies, is a form of control which incorporates the prediction of a system behaviour into its formulation. The prediction serves to estimate the future values of a variable based on the available system information. The more representative the information is of the system, the better is the accuracy of the prediction. The estimate of the future system variables can then be used in the design of control laws to achieve a good control performance, which is usually to drive or maintain the output to a desired set-point. This class of control methods which incorporates information or assumptions pertaining to the future values of the system output are generally referred to as *predictive control.*

Predictive control is usually considered when a better performance than that achievable by non-predictive control is required. This includes systems whose future behaviour can be quite different from that perceived of the present one, such as time-delay systems, high-order systems and poorly damped and non-minimum phase systems.

An early predictive control method was suggested by Smith. In 1959, Smith introduced the idea of using a predictor to overcome the problem encountered in controlling a system with dead time (Smith, 1959). The structure of the Smith-predictor controller (SPC) is shown in Fig. 1.1.

Consider a single-input and single-output process with delay in control described by

$$\begin{cases} \dot{x}(t) = Ax(t) + Bu(t-L), \\ \quad\;\; y(t) = Cx(t), \end{cases} \tag{1.1}$$

where u is the input, x is the $n \times 1$ state variable, y is the output, L is the delay. A, B, and C are constant matrices of appropriate dimensions for the process. Assume that a model of the process $G_p(s)$ is available which is described by:

$$G_p(s) = G(s)e^{-Ls},$$

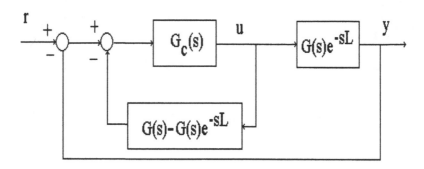

Fig. 1.1. The Smith-predictor controller

where $G(s) = C(sI - A)^{-1}B$ is a delay-free rational function.

The transfer function from the reference input r to the output y is given by:

$$G_{yr}(s) = \frac{y(s)}{r(s)} = \frac{G_c(s)G(s)}{1 + G_c(s)G(s)} e^{-L_s}.$$

This implies that the characteristic equation is free of the delay so that the primary controller $G_c(s)$ can be designed with respect to $G(s)$. This is the major advantage of the Smith-predictor control. The achievable performance can thus be greatly improved over a conventional single-loop system without the delay-free output prediction.

However, the Smith-predictor cannot be applied to unstable and poorly damped systems because the closed-loop poles always include the open-loop ones. Manitius and Olbrot (1979) investigated a finite spectrum assignment (FSA) control scheme which is applicable to systems with delay, including unstable systems. The FSA control law is given by:

$$u(t) = fe^{-AL}x(t + L). \tag{1.2}$$

Assuming a system model is available, $x(t + L)$ may be predicted from the prevailing input and state variables:

$$x(t + L) = e^{AL}x(t) + \int_{t-L}^{t} e^{A(t-\tau)}Bu(\tau)d\tau. \tag{1.3}$$

Thus, the control law becomes:

$$u(t) = fx(t) + f \int_{t-L}^{t} e^{A(t-L-\tau)} Bu(\tau) d\tau. \qquad (1.4)$$

The resulting closed-loop system can achieve desired response to arbitrary initial conditions.

For discrete-time systems, the future evolution of system output is first used in the filtering and prediction theory (Kalman, 1960). Later, Åström (1970) considered the use of predictors in discrete-time minimum variance control; in particular, he pioneered the polynomial approach to the design of predictors. In addition, in the work of Åström (1970), quality of control also largely depends on the performance criterion to be optimised, which is normally a single-step cost function.

Since the late 1970s, by incorporating a multistep cost function, a number of long-range predictive controllers have been described (Richalet et al., 1978; Peterka, 1984; Clarke and Mohtadi, 1989; Clarke and Scattolini, 1991; De Keyser and Van Cauwenberghe, 1981; Cutler and Ramaker, 1980; Ydstie, 1984; Clarke et al., 1987; Li et al., 1989; Scattolini and Bittanti, 1990; Clarke and Scattolini, 1991; Kouvaritakis et al., 1992; Lee et al., 1992; Yoon and Clarke, 1993; Lee and Yu, 1994; Lee et al., 1994; Scokaert, 1997; Camacho and Bordons, 1999). Due to its very desirable properties, long-range predictive control has increasingly been applied to industrial control problems. These approaches are based on a future control input sequence that optimises an open-loop *performance index*, according to a prediction of the system evolution from the current time t. Then, the sequence is actually applied to the system, until another sequence based on more recent data is newly computed. Among the well-known approaches to the design of long-range predictive controllers are Richalet's method (Richalet et al., 1978), also called the model algorithmic control (MAC), Culter's method (Cutler and Ramaker, 1980), also called the dynamic matrix control (DMC), De Keyser's method (De Keyser and Van Cauwenberghe, 1981), also called the extended prediction self-adaptive control (EPSAC), Ydstie's method (Ydstie, 1984), also called the extended horizon adaptive control (EHAC), Clarke's method (Clarke et al., 1987), also called the generalised predictive control (GPC), Clarke and Scattolini's method (Clarke and Scattolini, 1991; Rawlings and Muske, 1993; Rossiter and Kouvaritakis, 1993), also called the constrained receding-horizon predictive control (CRHPC), Kouvaritakis's method (Kouvaritakis et al., 1992), also called the stable generalized predictive control (SGPC), Scokaert's method (Scokaert, 1997), also called the infinite-horizon generalized predictive control (IHGPC), and Lee and Yu's method (Lee and Yu,1997; Casavola et al., 2000) or the min-max predictive control (MMPC). This book will mainly concentrate on the development and application of the GPC method of Clarke et al. (1987), since it is a general form of predictive control, and is simple, effective, and rather popularly used in many industrial applications.

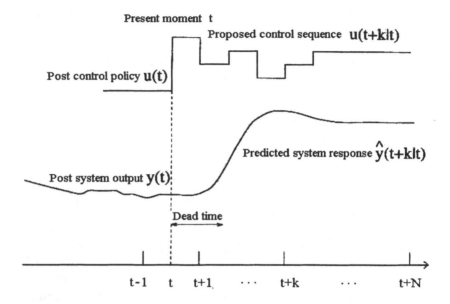

Fig. 1.2. Long-range predictive control

The GPC scheme is based on the principle of Fig. 1.2 and is characterised by the following steps:

- At the "present moment" t, a forecast of the process output over a long-range time horizon is made. This forecast is made implicitly or explicitly in the control algorithm and is based on a mathematical model of the system dynamics. Moreover, it is a function of the future control scenario. In Fig. 1.2, this future control scenario is presented as being constant over the prediction horizon. This is done for simplicity and is by no means a restriction.

- The control strategy is selected which brings the *predicted* system output back to the set-point in the "best" way according to a specific control objective.

- The resulting best candidate is then applied as a control action to the real system input but *only* at the present moment. At the next sampling instant the whole procedure is repeated, leading to an updated control action with corrections based on the latest measurements.

In order to further understand how the GPC works, a simple example is provided as follows:

Consider here the GPC control of a system:

$$y(t) = 1.7y(t-1) - 0.7y(t-2) + 0.9\Delta u(t-1) - 0.6\Delta u(t-2). \quad (1.5)$$

To predict the process output at each sampling instant, the model (1.5) is used in the following form as a predictive model:

$$\hat{y}(t+i/t) = 1.7y(t+i-1) - 0.7y(t+i-2) + 0.9\Delta u(t+i-1)$$
$$-0.6\Delta u(t+i-2), \ i \geq 1. \quad (1.6)$$

Let the prediction horizon and control horizon be selected as $p_1 = 3$ and $p_2 = 3$, respectively.

A range of three predictions are therefore needed. These may be derived by back substitution into the predictive model (1.6) as follows:

$$\hat{y}(t+1/t) = 1.7y(t) - 0.7y(t-1) + 0.9\Delta u(t) - 0.6\Delta u(t-1)$$
$$\hat{y}(t+2/t) = 1.7\hat{y}(t+1/t) - 0.7y(t) + 0.9\Delta u(t+1) - 0.6\Delta u(t)$$
$$= 2.19y(t) - 1.19y(t-1) + 0.9\Delta u(t+1) + 0.93\Delta u(t)$$
$$-1.02\Delta u(t-1)$$
$$\hat{y}(t+3/t) = 1.7\hat{y}(t+2/t) - 0.7\hat{y}(t+1/t) + 0.9\Delta u(t+2)$$
$$-0.6\Delta u(t+1)$$
$$= 2.533y(t) - 1.533y(t-1) + 0.9u(t+2) + 0.93\Delta u(t+1)$$
$$+0.951\Delta u(t) - 1.314\Delta u(t-1). \quad (1.7)$$

This yields the form of (1.8)

$$\begin{bmatrix} \hat{y}(t+1/t) \\ \hat{y}(t+2/t) \\ \hat{y}(t+3/t) \end{bmatrix} = \begin{bmatrix} 0.9 & 0 & 0 \\ 0.93 & 0.9 & 0 \\ 0.951 & 0.93 & 0.9 \end{bmatrix} \begin{bmatrix} \Delta u(t) \\ \Delta u(t+1) \\ \Delta u(t+2) \end{bmatrix}$$
$$+ \begin{bmatrix} 1.7y(t) - 0.7y(t-1) - 0.6\Delta u(t-1) \\ 2.19y(t) - 1.19y(t-1) - 1.02\Delta u(t-1) \\ 2.533y(t) - 1.533y(t-1) - 1.314\Delta u(t-1) \end{bmatrix}. \quad (1.8)$$

Note that the values of $y(t), y(t-1)$, and $\Delta u(t-1)$ are available. The GPC controller is obtained by minimising the following criterion:

$$J = \sum_{i=1}^{3} \{[y(t+i/t) - y_d(t+i)]^2 + 0.1[\Delta u(t+i-1)]^2\}, \quad (1.9)$$

where the signal $y_d(t+i)$ is the reference trajectory which is the desired system output. This minimisation process produces $\Delta u(t), \Delta u(t+1), \Delta u(t+2)$ as the following vector form:

$$
\begin{bmatrix} \Delta u(t) \\ \Delta u(t+1) \\ \Delta u(t+2) \end{bmatrix} = \begin{bmatrix} 0.8947 & 0.0929 & 0.0095 \\ -0.8316 & 0.8091 & 0.0929 \\ -0.0766 & -0.8316 & 0.8947 \end{bmatrix} \left\{ \begin{bmatrix} y_d(t+1) \\ y_d(t+2) \\ y_d(t+3) \end{bmatrix} \right.
$$
$$
\left. - \begin{bmatrix} 1.7y(t) - 0.7y(t-1) - 0.6\Delta u(t-1) \\ 2.19y(t) - 1.19y(t-1) - 1.02\Delta u(t-1) \\ 2.533y(t) - 1.533y(t-1) - 1.314\Delta u(t-1) \end{bmatrix} \right\}.
$$

This equation yields the future control increments for times t to $t+2$ as an open-loop strategy based upon information available at time t. The mechanism utilised for closing the loop and forcing a feedback control in GPC is to implement only the first element of $[\Delta u(t), \Delta u(t+1), \Delta u(t+2)]^T$, i.e., $\Delta u(t)$, and then to recompute the solution to the optimal control again for the next step using data available at time $t+1$. Since the model is linear time-invariant, carrying through the calculation of the first row above yields a time-invariant GPC control law:

$$
\Delta u(t) = -0.644\Delta u(t-1) + 1.7483y(t) - 0.7513y(t-1)
$$
$$
+0.8947y_d(t+1) + 0.0929y_d(t+2) + 0.0095y_d(t+3). \quad (1.10)
$$

This shows that GPC provides a way of synthesising a linear feedback controller by means of an optimisation criterion instead of, say, a pole placement design. An important feature of the GPC controller is that it has a natural structure of feedforward + feedback control.

Predictive control is an open methodology. That is, within the framework of predictive control, there are many ways to design a GPC-based controller. The goal of this book is to apply the concept of predictive control to a variety of control problems and to analyse their performance in terms of stability and robustness.

There is no standard notation to cover all the topics covered in this book. The most familiar notation from the literature is used whenever possible, but the overriding concern throughout the book has been consistent, to ensure that the reader can follow the ideas and the techniques through from one chapter to another.

Notation

$\text{diag}[x_1, x_2, ..., x_n]$ diagonal matrix with diagonal elements equal to $x_1, x_2, ...,$ x_n

$(\cdot)^T$ transpose of $(\cdot)$

$|(\cdot)|$ absolute value of $(\cdot)$

$\|M\|_l$ $l-$norm of M

$\|M\|_\infty$ infinity norm of M

$\lambda(\Lambda)$ denotes any eigenvalue of matrix Λ

$\lambda_{max}(\Lambda)$ the largest eigenvalue of Λ

$\lambda_{min}(\Lambda)$ the smallest eigenvalue of Λ

Mathematical Terminology.

Let A and B be statements. The following expressions are equivalent:

$$A \Rightarrow B,$$

if A holds, then deduce the result B.

The remaining notation, special terminology and abbreviations is defined in the main text.

CHAPTER 2
DEAD TIME COMPENSATORS

Dead time compensators were among the first predictive controllers formulated. This is not surprising since for time-delay systems, prediction of the behaviour of the system in the future is necessary to formulate the appropriate control action. For a non-predictive controller, the control action or the effect of control action cannot be realised until some time in the future when the system variables may have changed considerably such that the control action is no longer appropriate. While the proportional-integral-derivative (PID) controller is widely used in industry, its performance when used in time-delay systems is known to be limited. If a time-delay system is controlled by a PID controller, the derivative action is often disabled to reduce excessive overshoot and oscillations, due largely to misinterpretation of the non-responsiveness of the system (Hägglund and Åström, 1991). On the other hand, with no derivative action, there is no predictive capability in the remaining PI control action, which is precisely what is needed for these systems. Dead time compensators are employed when better performance is demanded.

In this chapter, two classical approaches to dead time compensation are described: the Smith-predictor controller (SPC) and finite spectrum assignment (FSA)

2.1 Smith-predictor Controller

The poor performance associated with the control of time-delay systems has been the subject of research for more than fifty years. It is realised that when the time constant is small compared to the time delay, the performance of the system is limited by the small loop gain necessary to preserve the stability of the system. Thus, when a better control performance is necessary, some form of dead time compensation becomes necessary. One of the most popular dead time compensation schemes is the SPC (Smith, 1959). However, the scheme is only applicable to stable systems, since the closed-loop poles will also include the open-loop ones, although variants of the basic scheme have been developed to expand the application of the SPC. This section describes the basic structure of the SPC. One of the main obstacles impeding the use of the SPC is the need for an accurate system model, which may be difficult to obtain in practice. Stability analysis of the SPC under modelling errors is also provided.

2.1.1 The Basic Smith Control System

Among the many dead time compensation control strategies available, the SPC strategy probably has the highest level of acceptance by practising engineers, and is consequently the most commonly used (Murill, 1988). It incorporates a model of the system, and is thus able to predict the output of the system. This allows the controller to be designed as though the system is delay-free, thus retaining the simple tuning features of PID controllers. The SPC offers potential improvement in the closed-loop performance over conventional controllers (Palmor and Shinnar, 1981; Tan et al., 1999; Tan et al., 2001), and it has already been extended to multivariable systems (Alevisakis and Seborg, 1974). The advantages of the SPC technique have been fully reported by Marshall et al. (1992).

The SPC was proposed by Smith et al. (1959) for dead time compensation and is shown in Fig. 2.1. The controller incorporates a model of the system, thus allowing for prediction of the system variables, and the controller may then be designed as though the system is delay-free.

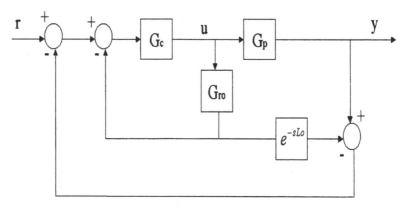

Fig. 2.1. The Smith-predictor controller

Assume that a model of the system $G_p(s)$ is available which is described by

$$G_{po}(s) = G_{ro}(s)e^{-L_o s},$$

where $G_{ro}(s)$ is a delay-free rational function.

The structure of the SPC is shown in Fig. 2.1. It can be shown that the closed-loop transfer function between the set-point and output is given by

$$G_{yr}(s) = \frac{G_c(s)G_p(s)}{1 + G_c(s)(G_{ro}(s) - G_{ro}(s)e^{-sL_o} + G_p(s))}. \tag{2.1}$$

In the case of perfect modelling, i.e., $G_{po}(s) = G_p(s)$, the closed-loop transfer function becomes

$$G_{yr}(s) = \frac{G_c(s)G_r(s)}{1 + G_c(s)G_r(s)}e^{-Ls}.$$

This implies that the characteristic equation is independent of the delay so that the primary controller $G_c(s)$ can be designed with respect to $G_r(s)$. The achievable performance can thus be greatly improved over a conventional single-loop system without delay-free output prediction.

2.1.2 Constraints of the SPC

The SPC is a very popular and very effective long dead time compensator for stable systems. The main advantage of the SPC method is that the time delay is eliminated from the characteristic equation of the closed-loop system. Despite the success of SPC in the process control industry, it fails in a very significant way in a few areas. First, the SPC is sensitive to modelling errors. Secondly, the SPC works well for set-point changes, but its performance in regulating against disturbances can be rather limited. Finally, the SPC cannot stabilise an unstable system. Variants of the SPC have evolved to try to address these concerns over the years. The next section discusses the stability properties of the SPC.

2.1.3 Robust Stability

The SPC is a model-based control system, and as such it will be subject to a modelling sensitivity problem when the model used in the primary controller design deviates from the actual system. Stability results are readily available for single-input/single-output (SISO) feedback control and *Internal Model Control* (IMC) design (Morari and Zafiriou, 1989). To consider the stability of the SPC, it can be arranged into an equivalent IMC system (Fig. 2.2), where $Q = \frac{G_c}{(1+G_cG_{ro})}$. In the following sub-section, nominal stability and robust

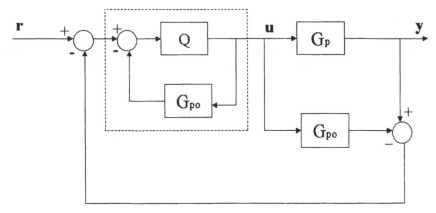

Fig. 2.2. IMC interpretation of the SPC

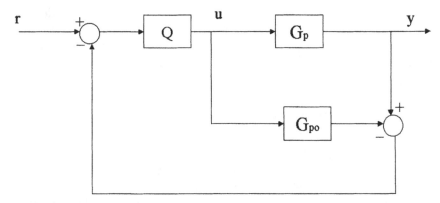

Fig. 2.3. Internal model control

stability of the SPC is investigated by referring to the existing results on SISO feedback control and IMC design.

To consider nominal stability , the SPC is arranged as shown in Fig. 2.2. This is equivalent to the IMC structure of Morari and Zafiriou (1989) shown in Fig. 2.3. Hence, it follows directly that the closed loop will be nominally stable if and only if both Q and G_{po} are stable, *i.e.*, the corresponding delay-free control system and the actual system G_p must be stable.

In the real world where the model does not represent the system exactly, *nominal stability* alone is not sufficient. The SPC is referred to as being *robustly stable* if it is designed such that the closed loop is stable for *all* members of a family of possible systems. In this section, the robust stability of the SPC of Fig. 2.1 will be investigated by referring to the existing results for SISO feedback control and IMC design (Fig. 2.3). The following assumptions are made.

Assumption 2.1

- The actual system $G_p(s)$ and the nominal system $G_{po}(s)$ do not have any unstable poles.

- The nominal closed-loop system $\tilde{G}_{yr}(s)$, between the set-point and output is stable.

The following is a useful lemma pertaining to SISO feedback control design (Morari and Zafiriou, 1989).

Lemma 2.1

Assume that the family of stable systems $\prod$ with norm-bounded uncertainty described by

$$\prod = \left\{ G_p : \left| \frac{G_p(j\omega) - G_{po}(j\omega)}{G_{po}(j\omega)} \right| = |l_m(j\omega)| \le \tilde{l}_m(\omega) \right\} \qquad (2.2)$$

have the same number of right-hand plane (RHP) poles and that a particular controller G_c stabilises a nominal plant G_p. Then, the closed loop is robustly stable with the controller G_c if and only if the complementary sensitivity function $\tilde{\eta}(s)$ for the nominal plant G_p satisfies the following bound:

$$\| \tilde{\eta}\tilde{l}_m \|_\infty \overset{\triangle}{=} \sup_{\omega} |\tilde{\eta}\tilde{l}_m(\omega)| < 1, \qquad (2.3)$$

where $\tilde{l}_m(\omega)$ is the bound on the multiplicative uncertainty $l_m(j\omega)$.

If the condition (2.3) is violated, then in the set $\prod$ defined by (2.2), there exists a plant G_p for which the closed loop with the controller G_c is unstable. If the set $\prod$ was obtained by approximating the true uncertainty regions with disks as described in (2.2), and therefore contains systems not present in the original uncertainty set, then (2.3) will be generally only sufficient for the original uncertainty set .

The following is a useful Lemma pertaining to IMC design (Morari and Zafiriou, 1989).

Lemma 2.2

Assume that the family of stable systems $\prod$ with norm-bounded uncertainty is described by (2.2) where $\tilde{l}_m(\omega)$ is the bound on the multiplicative uncertainty $l_m(j\omega)$. Then the IMC system of Fig. 2.2 (where Q is stable) is robustly stable if and only if

$$|Q(j\omega)G_{po}(j\omega)|\,\tilde{l}_m(\omega) < 1 \quad, \quad \forall\omega. \tag{2.4}$$

The SPC can be arranged into an equivalent IMC system in Fig. 2.2 where $Q = \frac{G_c}{(1+G_cG_{ro})}$. Note that, $\tilde{G}_{yr}(s) = Q(s)G_{po}(s)$ is the nominal closed-loop transfer function between the set-point and output, and under Assumption 2.1, Q is stable.

Theorem 2.1 is thus obtained which provides a necessary and sufficient condition for the robust stability of the SPC.

Theorem 2.1

Assume that the family of stable systems $\prod$ with norm-bounded uncertainty is described by (2.2). Then under Assumption 2.1, the SPC of Fig. 2.2 is robustly stable if and only if

$$\left|\tilde{G}_{yr}(j\omega)\right|\tilde{l}_m(\omega) < 1 \quad, \quad \forall\omega. \tag{2.5}$$

Remark 2.1

It is necessary to clarify what it means that Theorem 2.1 is not only sufficient for robust stability but also necessary. If the condition (2.5) is violated, then in the set $\prod$ defined by (2.2), there exists a system G_p for which the closed loop is unstable. If the set $\prod$ was obtained by approximating the true uncertainty regions with disks as described in (2.2), and therefore contains systems not present in the original uncertainty set, then (2.5) will be generally only sufficient for the original uncertainty set.

Remark 2.2

(2.5) can be rewritten as

$$|\tilde{G}_{yr}(j\omega)| < \tilde{l}_m^{-1}(\omega), \; \forall\omega.$$

Given the uncertainty bound $\tilde{l}_m(\omega)$, the robust stability of the closed-loop SPC can thus be ensured by a proper design of the ideal closed-loop transfer function given by $\tilde{G}_{yr}(s)$ so that the magnitude frequency response of $|\tilde{G}_{yr}(j\omega)|$ lies below that of $\tilde{l}_m^{-1}(\omega)$, and (2.5) is satisfied.

2.1.4 Robust Performance

Robust stability is the minimum requirement a control system has to satisfy to be useful in a practical environment where model uncertainty is an important issue. However, robust stability alone is not enough. Even if (2.5) is satisfied for the family $\prod$, there will exist a "worse case" system in $\prod$ for which the closed-loop system is on the verge of instability and for which the performance is arbitrarily poor. Thus, it is necessary to ensure that some performance specifications are met for *all* plants in the family $\prod$. Performance specifications stated in the H_∞ framework (Morari and Zafiriou, 1989) require

$$\max_{G_p \in \Pi} \| SW \|_\infty = \max_{G_p \in \Pi} \sup_\omega |S(j\omega)W(j\omega)| < 1, \qquad (2.6)$$

where S is the sensitivity function, and W is the performance weight. In general, W^{-1} provides a bound on the maximum peak of the function S, and imposes a minimum bandwidth constraint on the closed loop. For the simple choice of $W^{-1} = MP$, robust performance only requires the maximum peak of the sensitivity function to be less than MP with no bandwidth constraint. The reader may refer to Laughlin *et al.* (1987), for more details on the choice of the performance weight, W. As in Section 2.1.3, an existing result in IMC design (Morari and Zafiriou, 1989) will be applied to deduce results for the SPC.

Lemma 2.3 (Morari and Zafiriou, 1989)

Assume that the family of stable systems $\prod$ with norm-bounded uncertainty is described by (2.2). The IMC system of Fig. 2.3 (where Q is stable) will meet the performance specification (2.6) if and only if

$$|Q(j\omega)G_{po}(j\omega)|\tilde{l}_m(\omega) + |(1 - Q(j\omega)G_{po}(j\omega))W(j\omega)| < 1, \forall \omega.$$

Using Lemma 2.3 and the IMC interpretation of the SPC, the following theorem is obtained which provides a necessary and sufficient condition for robust performance of the SPC.

Theorem 2.2

Assume that the family of stable systems $\prod$ is described by (2.2). Then under Assumption 2.1, the SPC of Fig. 2.1 will meet the performance specification (2.6) if and only if

$$\left|\tilde{G}_{yr}(j\omega)\right|\tilde{l}_m(\omega) + \left|(1 - \tilde{G}_{yr}(j\omega))W(j\omega)\right| < 1 , \quad \forall \omega. \qquad (2.7)$$

Remark 2.3

(2.7) is necessary and sufficient for the norm-bounded uncertainty described in (2.2). In general, however, it is sufficient only when the true uncertainty region is only a portion of the uncertainty region described in (2.2).

Remark 2.4

(2.7) can also be rewritten as

$$\frac{|\tilde{G}_{yr}(j\omega)|}{1 - |(1 - \tilde{G}_{yr}(j\omega))W(j\omega)|} < \tilde{l}_m^{-1}(\omega).$$

Given the uncertainty bound $\tilde{l}_m(\omega)$ and the performance weight $W(j\omega)$, the robust performance design is thus to design a proper ideal closed-loop transfer function $\tilde{G}_{yr}(s)$ so that the magnitude–frequency response of

$$\frac{|\tilde{G}_{yr}(j\omega)|}{1 - |(1 - \tilde{G}_{yr}(j\omega))W(j\omega)|},$$

lies below that of $\tilde{l}_m^{-1}(\omega)$ to satisfy (2.7). Depending on $\tilde{l}_m(\omega)$ and the specification $W(j\omega)$, there may not be any $\tilde{G}_{yr}(j\omega)$ which satisfy (2.7), then $W(j\omega)$ may be too tight for the uncertainty present, and may have to be relaxed.

2.2 Finite Spectrum Assignment Control

The FSA originated with Manitius and Olbrot (1979) in the time domain. The frequency domain version was proposed by Ichikawa (1985), but the original version can be applied only to single-variable systems with distinct poles. Subsequent versions of the algorithm extend the application of FSA to systems with multiple poles and to asymptotic tracking and regulation. Unlike the SPC, the FSA can arbitrarily assign the closed-loop poles and therefore can be applied to poorly damped and unstable systems. Like the SPC, the structure of the FSA is physically realisable and implementation can be as simple and efficient. One weakness of FSA, compared with SPC, is probably that many developments in FSA only recently been reported and accessibility in terms of books and literature is not as extensive as SPC. Therefore, control engineers may not be as familiar and comfortable with the concept of FSA. This section gives a concise coverage of the essentials of FSA.

2.2.1 Finite Spectrum Assignment (FSA) Algorithms

The FSA algorithm for delay systems due to Ichikawa (1985) is actually
an extension of Wolovich's frequency domain pole assignment for delay-free
systems (Wolovich, 1974). It is thus helpful to briefly review the latter first.

Let a delay-free system be described by

$$Y(s) = G(s)U(s) = \frac{a(s)}{b(s)}U(s), \qquad (2.8)$$

where $a(s)$ and $b(s)$ are polynomials of degree m and n respectively, $0 \leq m \leq n - 1$, and $a(s)$ and $b(s)$ are assumed to be coprime. The system may
be unstable and/or of non-minimum phase. Pole assignment is a means of
stabilising an unstable system.

Denote the n-degree monic polynomial which is the characteristic polynomial
of the closed-loop system by $p(s)$. Denote $b(s) - p(s)$ by $f(s)$, which is of
degree $n - 1$ at most. Introduce any $(n - 1)$-degree monic asymptotically
stable polynomial $q(s)$ (in what follows the term asymptotic will be omitted).
The physical meaning of $q(s)$ is that it is the characteristic polynomial of a
reduced-order Luenberger observer that yields a state estimate $\hat{x}(t)$ from
the available $u(t)$ and $y(t)$, but the observer is constructed implicitly in the
frequency pole assignment. Consider a polynomial equation:

$$k(s)b(s) + h(s)a(s) = q(s)f(s), \qquad (2.9)$$

where $k(s)$ and $h(s)$ are unknown polynomials. Equation (2.9) yields a unique
solution for $k(s)$ and $h(s)$ of degree at most $n - 2$ or $n - 1$ respectively. Using
the solution $k(s)$ and $h(s)$, the following control law is constructed:

$$U(s) = \frac{k(s)}{q(s)}U(s) + \frac{h(s)}{q(s)}Y(s) + R(s), \qquad (2.10)$$

where $r(t)$ is an external reference input. The control law (2.10) achieves the
desired pole assignment of $p(s)q(s)$ since (2.8)–(2.10) imply that

$$p(s)q(s)U(s) = b(s)q(s)R(s).$$

The control law (2.10) can be written in the time domain as

$$q(\mathbf{D})u(t) = k(\mathbf{D})u(t) + h(\mathbf{D})y(t) + q(\mathbf{D})r(t), \qquad (2.11)$$

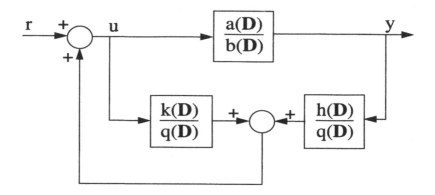

Fig. 2.4. Pole assignment control system for undelayed system

Consider now a time-delay system:

$$Y(s) = G(s)U(s) = \frac{a(s)}{b(s)}e^{-Ls}U(s), \qquad (2.12)$$

where $a(s)/b(s)$ is a strictly proper and coprime rational function of order n with $b(s)$ *monic*. $L > 0$ is the dead time, $b(s)$ is assumed to have no multiple zeros. The system with delay is of infinite dimension, and the whole state of the system is considered as the set of $x(t)$, the state of the lumped portion, and the time function $u(\tau)$, $t - L \leq \tau \leq t$. Naturally, pole assignment will require feedback of the whole state with appropriate coefficients. The state-space approach developed by Furukawa and Shimemura (1983) used another idea to feed back the predicted state $x(t + L)$. Since the predicted state, however, consists of $x(t)$ and $u(\tau)$, $t - L \leq \tau \leq t$, both ideas are considered to be equivalent.

The approach to be presented in this section depends on the former idea. The system state $x(t)$ can be estimated from $u(t - L)$ and $y(t)$, so that the first term in the control law (2.11) is changed to $k(\mathbf{D})u(t - L)$. In order to obtain the same result as was obtained by the time-domain method, some trial and error was needed in the determination of appropriate feedback coefficients for $u(\tau)$. Because $u(\tau)$ is infinite-dimensional, coefficients assume the form of a time function. Let the required closed-loop *monic* polynomial of degree n and the observer polynomial of degree $(n-1)$, respectively, be $p(s)$ and $q(s)$. Define

$$f(s) := b(s) - p(s), \tag{2.13}$$

and consider the fraction $f(s)p^{-1}(s)$. Since $f(s)$ is of degree $n - 1$ at most, $p(s)$ is of degree n, and $p(s)$ is assumed to have no multiple zeros, then the partial fraction expansion of $f(s)p^{-1}(s)$ gives

$$\frac{f(s)}{b(s)} = \sum_{i=1}^{n} \frac{c_i}{s - \lambda_i}, \tag{2.14}$$

where $\lambda_i, i = 1, 2, \cdots, n$, are n distinct poles of the system.

Define another polynomial $f_L(s)$ by

$$f_L(s) := b(s) \sum_{i=1}^{n} \frac{c_i e^{\lambda_i L}}{s - \lambda_i}, \tag{2.15}$$

so that the degree of $f_L(s)$ is at most $(n - 1)$. One solves the following polynomial equation:

$$k(s)b(s) + h(s)a(s) = q(s)f_L(s), \tag{2.16}$$

for $k(s)$ and $h(s)$ such that $k(s)/q(s)$ and $h(s)/q(s)$ are both proper. The control $u(t)$ is obtained from

$$q(\mathbf{D})u(t) = k(\mathbf{D})u(t - L) + h(\mathbf{D})y(t)$$
$$+q(\mathbf{D}) \int_{-L}^{o} \sum_{i=1}^{n} c_i e^{-\lambda_i \tau} u(t + \tau) d\tau + q(\mathbf{D})r(t), \tag{2.17}$$

where r is the set-point.

Theorem 2.3

The control law (2.17) is realisable and achieves arbitrary finite spectrum assignment.

Proof

Since the polynomials $k(s)$ and $h(s)$ are of degree $n - 2$ and $n - 1$ (respectively) at most, both $k(s)q^{-1}(s)$ and $h(s)q^{-1}(s)$ are realisable. Furthermore, since $u(t + \tau)$, $-L \leq \tau \leq 0$, is the past history of the control signal over the finite interval L, and $c_i e^{-\lambda_i \tau}$ is definite over the interval, the integral

$\int_{-L}^{0} c_i e^{-\lambda_i \tau} u(t+\tau) d\tau)$ can be determined for any t. Therefore, the control law (2.17) is realisable in theory, although computation of the integral in practice requires some time.

Taking the Laplace transform of (2.17) yields:

$$q(s)U(s) = k(s)U(s)e^{-Ls} + h(s)Y(s)$$
$$+ q(s) \int_{-L}^{0} \sum_{i=1}^{n} c_i e^{-\lambda_i \tau} U(s)e^{\tau s} d\tau + q(s)R(s).$$

However, one notes that

$$\int_{-L}^{0} \sum_{i=1}^{n} c_i e^{-\lambda_i \tau} U(s)e^{\tau s} d\tau = \sum_{i=1}^{n} c_i \int_{-L}^{0} e^{(s-\lambda_i)\tau} d\tau U(s)$$

$$= \sum_{i=1}^{n} \frac{c_i}{s-\lambda_i} U(s) - \sum_{i=1}^{n} \frac{c_i}{s-\lambda_i} e^{\lambda_i L} U(s)e^{-Ls}$$

$$= \frac{f(s)}{b(s)} U(s) - \frac{f_L(s)}{b(s)} U(s)e^{-Ls}.$$

Therefore, it follows that

$$q(s)U(s) = k(s)U(s)e^{-Ls} + h(s)Y(s) + \frac{q(s)f(s)}{b(s)} U(s) \qquad (2.18)$$

$$- \frac{q(s)f_L(s)}{b(s)} U(s)e^{-Ls} + q(s)R(s).$$

On the other hand, multiplying both sides of (2.16) by $b^{-1}(s)U(s)e^{-Ls}$ yields:

$$k(s)U(s)e^{-Ls} + h(s)Y(s) = \frac{q(s)f_L(s)}{b(s)} U(s)e^{-Ls}. \qquad (2.19)$$

It then follows from (2.18) and (2.19) that

$$q(s)U(s) = \frac{q(s)f(s)}{b(s)} U(s) + q(s)R(s),$$

or

$$q(s)p(s)U(s) = b(s)q(s)R(s),$$

so that the closed-loop poles are determined by $p(s)$ and $q(s)$ and they can be assigned arbitrarily.

One also sees that

$$Y(s) = \frac{a(s)}{b(s)}e^{-Ls}U(s) = \frac{q(s)a(s)}{q(s)p(s)}e^{-Ls}R(s).$$

However, since $q(s)$ is stable, the above reduces to

$$Y(s) = \frac{a(s)}{p(s)}e^{-Ls}R(s). \tag{2.20}$$

The FSA scheme is shown in Fig. 2.5. It is interesting to note that although (2.18) is derived from (2.17), the control law must not be represented in the form of (2.18), because the third and fourth terms of (2.18) bring into the control system unnecessary dynamics, destabilising the closed loop when the system is unstable.

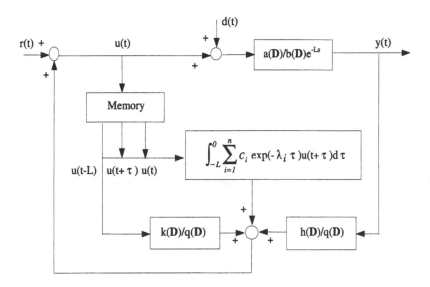

Fig. 2.5. The FSA system

2.2.2 Tracking and Regulation

Asymptotic tracking and regulation have always been desirable properties for any process control systems. It is, however, noted that a non-zero steady state error may exist even when an integrator is cascaded with the system and the FSA is applied to the cascaded system in the usual way.

Example 2.1

Consider the system:

$$G_p(s) = \frac{1}{5s+1}e^{-10s},$$

which is cascaded with an integrator so that the generalised system is

$$G(s) = \frac{1}{s(5s+1)}e^{-10s}.$$

To design an FSA for a delay system, a $p(s)$ may be chosen such that $p(0) = a(0)$ so that from (2.20), the closed-loop transfer function,

$$G_{yr}(s) = \frac{a(s)}{p(s)}e^{-Ls},$$

has a static gain of 1 to achieve asymptotic tracking. According to (2.12), $a(s) = \frac{1}{5}$ and $a(0) = 0.2$, $p(s)$ is then specified as

$$p(s) = s^2 + 0.894s + 0.2,$$

with a damping factor of 1, and the resultant closed-loop transfer function is

$$G_{yr}(s) = \frac{0.2}{s^2 + 0.894s + 0.2}e^{-10s},$$

and has no steady-state error in response to a step set-point change. However, the static gain cannot be guaranteed to be 1 if there is a perturbation in the system. For instance, if $a(s)$ is changed from 0.2 to 0.18, the transfer function becomes

$$\tilde{G}_{yr}(s) = \frac{0.18}{s^2 + 0.894s + 0.2}e^{-10s}.$$

The closed-loop system then has a steady state error. This is shown in Fig. 2.6. Therefore, the incorporation of an integrator in FSA systems usually cannot remove offset because of inevitable modelling errors.

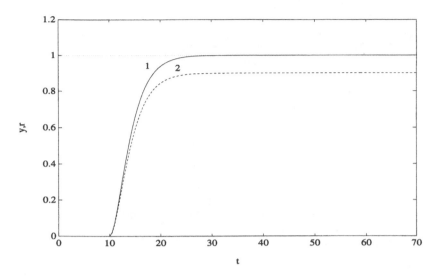

Fig. 2.6. Performance of the FSA system for (1) the original system, (2) the system after a parameter change

In order to achieve asymptotic tracking and regulation in the presence of possible modelling errors, a modified FSA scheme is shown in Fig. 2.7. Basically, the error between the set-point and system output is formed after the system and this is followed by an integrator. The resultant generalised system is then stabilised using the ordinary FSA method.

Let a system be described by

$$Y(s) = G_p(s)U(s) = \frac{a_p(s)}{b_p(s)}e^{-Ls}U(s), \qquad (2.21)$$

where $a_p(s)/b_p(s)$ is a proper and coprime rational function with $b_p(s)$ monic. The generalised system consisting of the system and an integrator is represented by the transfer function:

$$G(s) = \frac{a(s)}{b(s)}e^{-Ls}, \qquad (2.22)$$

where $a(s) = -a_p(s)$ and $b(s) = sb_p(s)$. Note that in the general case, $G(s)$ refers to the actual system. However, for the modified FSA method, the actual system is denoted by $G_p(s)$ and the generalised system by $G(s)$.

It is assumed that the system has no zero at $s = 0$ so that $a(s)$ and $b(s)$ are coprime. Signals v and u in Fig. 2.7 are viewed as the system output and

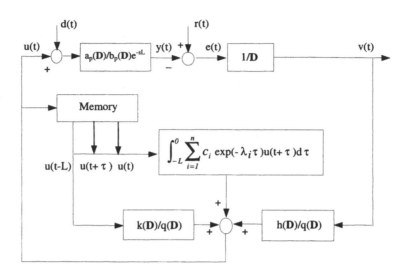

Fig. 2.7. The modified FSA system

input with the transfer function $G(s)$ given by (2.22), and we define $f(s)$, $f_L(s)$, $k(s)$ and $h(s)$ as in (2.13)–(2.16). However, the control law is modified to

$$
\begin{aligned}
q(\mathbf{D})u(t) = \quad & k(\mathbf{D})u(t - L) + h(\mathbf{D})v(t) \\
& + q(\mathbf{D}) \int_{-L}^{o} \sum_{i=1}^{n} c_i e^{-\lambda_i \tau} u(t + \tau) d\tau.
\end{aligned} \tag{2.23}
$$

The differences between (2.17) and (2.23) are that the system output y in (2.17) is replaced with the generalised system output v in (2.23), and that the set-point r in (2.17) disappears in (2.23) as it now enters the system through the generalised system. Taking the Laplace transform of (2.23) yields

$$
q(s)U(s) = k(s)e^{-Ls}U(s) + h(s)V(s) + \frac{q(s)f(s)}{b(s)}U(s)
$$
$$
- \frac{q(s)f_L(s)}{b(s)}e^{-Ls}U(s). \tag{2.24}
$$

The closed-loop relationship is now given by

$$
Y(s) = \frac{-h(s)a(s)}{p(s)q(s)}e^{-Ls}R(s), \tag{2.25}
$$

and the nominal system is stable provided that $p(s)$ and $q(s)$ are Hurwitz polynomials .

Theorem 2.4

The system in Fig. 2.7 achieves asymptotic tracking and regulation in response to a step set-point change and/or load disturbance provided that it is stable.

Proof

With (2.13), (2.16), (2.21), (2.23), it follows from Fig. 2.7 that the error between the set-point and output is given by

$$E(s) := R(s) - Y(s) = G_{er}(s)R(s) + G_{ew}(s)W(s), \tag{2.26}$$

where

$$G_{er}(s) = \frac{q(s)p(s) + h(s)a(s)e^{-Ls}}{q(s)p(s)}, \tag{2.27}$$

and

$$G_{ew}(s) = -G_p(s)\frac{q(s)p(s) + h(s)a(s)e^{-Ls}}{q(s)p(s)}. \tag{2.28}$$

Define

$$\phi(s) = q(s)p(s) + h(s)a(s)e^{-Ls}. \tag{2.29}$$

It is now claimed that

$$\phi(\lambda_i) = 0, \quad i = 1, 2, \cdots, n, \tag{2.30}$$

that is, each zero λ_i of $b(s)$ is that of $\phi(s)$. Indeed, at $s = \lambda_i$, (2.16) becomes

$$h(\lambda_i)a(\lambda_i) = q(\lambda_i)f_L(\lambda_i).$$

It follows from (2.15) and (2.14) that

$$f_L(\lambda_i) = e^{\lambda_i L} \cdot c_i \frac{b(s)}{s - \lambda_i} \big|_{s=\lambda_i} = e^{\lambda_i L} f(\lambda_i).$$

Hence, it follows that

$$\phi(\lambda_i) = q(\lambda_i)p(\lambda_i) + q(\lambda_i)f(\lambda_i).$$

Furthermore, it follows from (2.13) that

$$\phi(\lambda_i) = q(\lambda_i)b(\lambda_i) = 0,$$

and thus (2.30) is true. For a step disturbance, $W(s) = \frac{\beta}{s}$, one sees from (2.28) and (2.29) that

$$G_{ew}(s)W(s) = \frac{\beta a(s)e^{-Ls}}{q(s)p(s)} \cdot \frac{\phi(s)}{b(s)}. \tag{2.31}$$

(2.30) implies that $\phi(s)$ cancels $b(s)$ completely and (2.31) is thus stable since both $q(s)$ and $p(s)$ are stable by design. Similarly, for a step set-point change, $R(s) = \frac{\alpha}{s}$, it follows from (2.27) that

$$G_{er}(s)R(s) = \frac{\alpha b_p(s)}{q(s)p(s)} \cdot \frac{\phi(s)}{b(s)}, \tag{2.32}$$

which is, again by (2.30), stable. The stability of (2.31) and (2.32) together implies that $E(s)$ in (2.26) is stable and $e(\infty) = 0$. The result follows.

Remark 2.5

If a set-point change or load disturbance is of a type other than a step, the integrator in Fig. 2.7 should be replaced with $1/\psi(s)$, where $\psi(s)$ is the least common denominator of set-point and disturbance signals in the Laplace domain. Asymptotic tracking and regulation are still achievable with the generalised system, $-G_p(s)/\psi(s)$, stabilised with the FSA scheme.

Remark 2.6

The modified FSA system above will maintain asymptotic tracking and regulation in the presence of a system perturbation provided that the perturbed system remains stable. Physically, when the system stability is preserved under perturbations in the system, all signals in Fig. 2.7 will become finite constants as the time approaches infinity. But, in order for the output of the integrator, v, to remain constant in the steady state, its input, e, must be zero, implying that asymptotic tracking and regulation have been achieved.

2.2.3 Practical Stability

Assume that $\hat{G}(s) = \frac{\hat{a}(s)}{\hat{b}(s)}e^{-\hat{L}s}$ is a model for the actual generalised system $G(s) = \frac{a(s)}{b(s)}e^{-Ls}$. $G(s)$ is unknown and only $\hat{G}(s)$ is available for FSA system design. In general, $\hat{G}(s) \neq G(s)$. For practical use, an FSA system must remain stable for small differences between $G(s)$ and $\hat{G}(s)$. It can be shown that the closed-loop transfer function is given by

$$Y(s) = -\frac{h(s)a(s)}{b(s)w(s)}e^{-sL}R(s), \tag{2.33}$$

where

$$w(s) := \frac{q(s)p(s)}{\hat{b}(s)} + \frac{h(s)\hat{a}(s)}{\hat{b}(s)}e^{-\hat{L}s} - \frac{h(s)a(s)}{b(s)}e^{-Ls}. \tag{2.34}$$

Denoting $\bar{w}(s) = b(s)w(s)$, the characteristic equation is thus given by

$$\bar{w}(s) = 0.$$

The system stability depends entirely on the location of the zeros of the characteristic equation. Let $\sum$ denote the set of real parts of all the zeros of $\bar{w}(s)$:

$$\sum := \{\sigma | \sigma = \text{Re}z, \bar{w}(z) = 0\}.$$

Noting that $\sum$ is in general an infinite set (Bellman and Cooke, 1963), stability of the system is then defined as follows:

Definition 2.1

The closed-loop system is said to be stable if and only if (iff) $\sum$ has a negative upper bound.

The nominal system with $G(s) = \hat{G}(s)$ is always designed such that it is stable. But if there is a perturbation in the process, the system will not necessarily preserve its stability. This is shown in the following example.

Example 2.2

Consider a system with the transfer function :

$$G(s) = \frac{\epsilon s^2 + \epsilon s + 1}{(\epsilon s + 1)(s + 1)} e^{-s},$$

where $\epsilon > 0$ is a small parameter. It is common practice to obtain the nominal process by neglecting the small parameter ϵ so that

$$\hat{G}(s) = \frac{1}{s+1} e^{-s}.$$

Solving the design equations (2.13)–(2.16) with stable $p(s)$ and $q(s)$ specified as

$$p(s) = s + \alpha, \quad \alpha > \epsilon + 1, \tag{2.35}$$

and

$$q(s) = 1,$$

gives

$$f(s) = 1 - \alpha, \quad \frac{f(s)}{\hat{b}(s)} = \frac{1-\alpha}{s+1},$$

$$f_L(s) = (1-\alpha)e^{-1}, \quad k(s) = 0,$$

and

$$h(s) = (1-\alpha)e^{-1}.$$

In this example, $p(s)$, $q(s)$ and $\hat{b}(s)$ are all stable, and $w(s)$ has the same zeros in the closed right-half complex plane as those of

$$\tilde{w}(s) := w(s) \frac{\hat{b}(s)}{p(s)q(s)} = 1 + \gamma(s)e^{-s},$$

where

$$\gamma(s) = \frac{\alpha - 1}{e} \cdot \frac{\epsilon s^2}{(\epsilon s + 1)(s + \alpha)}.$$

It can be shown (Bellman and Cooke, 1963) that the zeros of $\tilde{w}(s)$ make up an infinite chain asymptotic to those of the comparison function:

$$f_c(s) := 1 + \gamma(\infty)e^{-s}.$$

Since the chain of the zeros of $f_c(s)$ is located on the line Re $s = \log\gamma(\infty) = \log\frac{\alpha-1}{e} > 0$ by (2.35), $\tilde{w}(s)$ has an infinite number of zeros in the right-half plane no matter how small ϵ is. In other words, the closed-loop system is only stable when $\epsilon = 0$, and an infinitesimal perturbation of the system transfer function coefficients will destabilise a nominally stable system. Obviously, such a system is useless in industrial control and the stability of the system in the face of a mismatch has to be considered. The following definition is first made for practical stability.

Definition 2.2

Let a finite spectrum-assigned system with the nominal system $\hat{G}(s) = \frac{\hat{a}(s)}{\hat{b}(s)}e^{-\hat{L}s}$ be stable. The closed-loop system is said to be practically stable iff there exist positive numbers, ω_M and δ, such that the system is stable for each $G(s) = \frac{a(s)}{b(s)}e^{-Ls}U(s)$ satisfying

$$\left|\frac{G(j\omega)}{\hat{G}(j\omega)} - 1\right| < \delta, \quad 0 \le \omega \le \omega_M. \tag{2.36}$$

Theorem 2.5 deals with strictly proper systems.

Theorem 2.5

If $\frac{\hat{a}(s)}{\hat{b}(s)}$ is strictly proper and each $\frac{a(s)}{b(s)}$ is strictly proper and has the same number of unstable poles as $\frac{\hat{a}(s)}{\hat{b}(s)}$ has, then the closed-loop system is practically stable.

Proof

Denoting $Z^+(\bar{w})$ as the number of unstable zeros of the characteristic equation $\bar{w}(s) = 0$, it has to be shown that $Z^+(\bar{w}) = 0$ if the conditions of Theorem 2.5 hold. It is observed that

$$\bar{w}(s) = \frac{b(s)}{\hat{b}(s)} \cdot p(s)q(s)\tilde{w}(s), \tag{2.37}$$

where

$$\tilde{w}(s) := w(s)\frac{\hat{b}(s)}{q(s)p(s)} = 1 + \frac{\hat{b}(s)h(s)}{q(s)p(s)}\left[\frac{\hat{a}(s)}{\hat{b}(s)}e^{-\hat{L}s} - \frac{a(s)}{b(s)}e^{-Ls}\right].$$

By (2.34) and (2.37), $\bar{w}(s)$ has the following form:

$$\bar{w}(s) = b(s) \cdot \frac{\phi_o(s)}{\hat{b}(s)} - h(s)a(s)e^{-Ls}, \tag{2.38}$$

where

$$\phi_o(s) = q(s)p(s) + h(s)\hat{a}(s)e^{-\hat{L}s}.$$

It is now claimed that $\bar{w}(s)$ has no unstable poles, i.e.,

$$P^+(\bar{w}) = 0, \tag{2.39}$$

($P^+(\bar{w})$ is the number of unstable poles of $\bar{w}$) since it can be inferred that $\phi_o(s)$ has exactly the same zeros as $\hat{b}(s)$ has.

Denoting by $N(k, \bar{w})$, the net number of clockwise encirclements of the point $(k, 0)$ by the image of the Nyquist D contour under $\bar{w}$, and by the principle of argument, it follows that

$$N(0, \bar{w}) = Z^+(\bar{w}) - P^+(\bar{w}).$$

Since $P^+(\bar{w}) = 0$, $Z^+(\bar{w}) = 0$ if and only if $N(0, \bar{w}) = 0$, i.e., the Nyquist curve of $\bar{w}(s)$ does not encircle the origin. Since $b(s)$ has the same number of unstable zeros as $\hat{b}(s)$ has, and $p(s)$ and $q(s)$ are user-specified and Hurwitz polynomials, (2.37) implies that

$$N(0, \bar{w}) = N(0, \tilde{w}) = N(-1, \tilde{w} - 1).$$

For stability, it is necessary that

$$N(-1, \tilde{w} - 1) = 0.$$

$\tilde{w}(s)$ can be rewritten as

$$\tilde{w}(s) = 1 + \frac{\hat{a}(s)h(s)}{q(s)p(s)} \cdot e^{-\hat{L}s} \cdot \left[1 - \frac{G(s)}{\hat{G}(s)}\right], \tag{2.40}$$

or

$$\tilde{w}(s) - 1 = G_{yr}^*(s)l_m(s),$$

where $G_{yr}^*(s) = -\frac{\hat{a}(s)h(s)}{p(s)q(s)}e^{-s\hat{L}}$ is the ideal closed-loop transfer function between the set-point and system output.

Since $\frac{\hat{b}(s)}{p(s)}$ and $\frac{h(s)}{q(s)}$ are both proper, $\frac{\hat{a}(s)}{\hat{b}(s)}$ and $\frac{a(s)}{b(s)}$ are both strictly proper, it follows that for the part of the Nyquist D contour where $|s| \longrightarrow \infty$, and $\mathrm{Re}\, s \geq 0$,

$$|G_{yr}^*(s)l_m(s)| = \left|\frac{\hat{b}(s)h(s)}{p(s)q(s)}(\hat{G}(s) - G(s))\right|$$

$$\leq \left|\frac{\hat{b}(s)}{p(s)}\right| \cdot \left|\frac{h(s)}{q(s)}\right| \cdot \left[\left|\frac{\hat{a}(s)}{\hat{b}(s)}\right| \cdot \left|e^{-\hat{L}s}\right| + \left|\frac{a(s)}{b(s)}\right| \cdot \left|e^{-Ls}\right|\right] \longrightarrow 0. \qquad (2.41)$$

Hence, there exists a positive number ω_M such that

$$|\tilde{w}(j\omega) - 1| < 1, \quad \omega > \omega_M. \qquad (2.42)$$

The stability of $\frac{\hat{a}(s)h(s)}{q(s)p(s)} \cdot e^{-\hat{L}s}$ implies that

$$\left|\frac{\hat{a}(j\omega)h(j\omega)}{q(j\omega)p(j\omega)} \cdot e^{-j\hat{L}\omega}\right| < M, \quad 0 \leq \omega \leq \omega_M, \qquad (2.43)$$

for some positive M. Take a positive δ such that $M\delta < 1$, then for each system $G(s) = \frac{a(s)}{b(s)}e^{-Ls}$ satisfying

$$\left|\frac{G(j\omega)}{\hat{G}(j\omega)} - 1\right| < \delta, \quad 0 \leq \omega \leq \omega_M,$$

there holds

$$|\tilde{w}(j\omega) - 1| < 1, \quad 0 \leq \omega \leq \omega_M. \qquad (2.44)$$

The equations (2.42) and (2.44) together imply that

$$|\tilde{w}(s) - 1| < 1$$

for each point of the Nyquist Contour and thus the Nyquist curve of $\tilde{w}(s)$ does not encircle the origin and the proof is completed.

One notes that if, in addition to strict properness, $\frac{\hat{a}(s)}{\hat{b}(s)}$ and $\frac{a(s)}{b(s)}$ are stable, then the conditions of Theorem 2.5 are satisfied. The following corollary is obtained.

Corollary 2.1

If $\frac{\hat{a}(s)}{\hat{b}(s)}$ and $\frac{a(s)}{b(s)}$ are stable and strictly proper, then the closed-loop system is practically stable.

It may, however, happen that either $\frac{\hat{a}(s)}{\hat{b}(s)}$ or any $\frac{a(s)}{b(s)}$ is not strictly proper, then Theorem 2.5 does not apply. In this case, the following necessary and sufficient condition for practical stability can be established under certain assumptions on the systems.

Theorem 2.6

If $\frac{\hat{a}(s)}{\hat{b}(s)}$ is stable and minimum phase, and all $\frac{a(s)}{b(s)}$ are stable, then the closed-loop system is practically stable iff

$$\lim_{\omega \to \infty} (\kappa \omega^{\lambda} + 2)|G_{yr}^{*}(j\omega)| < 1, \tag{2.45}$$

where the non-negative real κ and non-negative integer λ satisfy

$$\left| \frac{a(j\omega)/b(j\omega)}{\hat{a}(j\omega)/\hat{b}(j\omega)} - 1 \right| \leq \kappa \omega^{\lambda}, \quad \omega > \beta \geq 0.$$

Proof

Under the given conditions, Theorem 1 of Yamanaka and Shimemura (1987) is applicable to $\tilde{w}(s)$ and the result follows directly.

2.2.4 Robust Stability

In this sub-section, a robust stability analysis for FSA will be provided.

Theorem 2.7

Assume that the family of systems $\prod$ with norm-bounded uncertainty is described by

$$\prod = \left\{ G : \left| \frac{\hat{G}(j\omega) - G(j\omega)}{\hat{G}(j\omega)} \right| = |l_m(j\omega)| \leq \tilde{l}_m(\omega) \right\}, \qquad (2.46)$$

where $G(s) = \frac{a(s)}{b(s)} e^{-Ls}$, $\hat{G}(s) = \frac{\hat{a}(s)}{\hat{b}(s)} e^{-\hat{L}s}$, and $\tilde{l}_m(\omega)$ is the bound on the multiplicative uncertainty $l_m(j\omega)$. Furthermore, both $\frac{a(s)}{b(s)}$ and $\frac{\hat{a}(s)}{\hat{b}(s)}$ are strictly proper and they have the same number of unstable poles. Then, the FSA system is robustly stable if

$$\left| G_{yr}^*(j\omega) \right| \tilde{l}_m(\omega) < 1 \ , \quad \forall \omega. \qquad (2.47)$$

Proof

Under the given conditions, the evaluation of (2.41) for the imaginary axis of the Nyquist Contour provides the robust stability condition, i.e.,

$$|\tilde{w}(j\omega) - 1| < 1, \quad -\infty < \omega < \infty. \qquad (2.48)$$

Given $\tilde{w}$ in (2.40) and the specified uncertainty bound in (2.46), Theorem 2.7 follows directly.

Remark 2.7

For robust stability, the amplitude response of $|G_{yr}^*(j\omega)|$ should be shaped to lie below that of $\tilde{l}_m^{-1}(\omega)$. This can be done through a proper assignment of the ideal closed-loop poles via the desired closed-loop polynomial $p(s)$ and the observer polynomial $q(s)$. The closed-loop zeros contributed by the polynomials $\hat{a}(s)$ and $h(s)$, however, cannot be arbitrarily assigned. The dynamics introduced by these zeros also has to be accounted for in robust stability.

2.2.5 Performance Robustness

In this sub-section, the robust performance of the FSA system will be investigated. The following is a result for the robust performance of the FSA system.

Theorem 2.8

Assume that the family of systems $\prod$ is described by (2.46). Furthermore, both $\frac{a(s)}{b(s)}$ and $\frac{\hat{a}(s)}{\hat{b}(s)}$ are strictly proper and they have the same number of unstable poles . Then, the FSA system will meet the performance specification of (2.6) if

$$\left|G_{yr}^*(j\omega)\right|\tilde{l}_m(\omega) + \left|(1 - G_{yr}^*(j\omega))W(j\omega)\right| < 1 \quad, \quad \forall\omega. \tag{2.49}$$

Proof

The sensitivity function S is defined as

$$S(s) = 1 - G_{yr}(s).$$

Hence, it can be shown from (2.33) that

$$S(s) = \frac{1 + \frac{h(s)\hat{a}(s)}{p(s)q(s)}e^{-s\hat{L}}}{1 + \frac{h(s)\hat{a}(s)}{p(s)q(s)}e^{-s\hat{L}}(1 - \frac{a\hat{b}}{\hat{a}b}e^{-s(L-\hat{L})})},$$

which can be further simplified to

$$S(s) = \frac{1 - G_{yr}^*(s)}{1 - G_{yr}^*(s)l_m(s)}.$$

Replacing s by $j\omega$,

$$|S(j\omega)W| = \left|\frac{(1 - G_{yr}^*(j\omega))W(j\omega)}{1 - G_{yr}^*(j\omega)l_m(j\omega)}\right|$$

Fig. 2.8 shows a boundary plane on which $G_{yr}^*l_m$ is located for all $G \in \prod$. For robust stability, it follows from (2.47) that $|G_{yr}^*(j\omega)|\tilde{l}_m(\omega) < 1$, $\forall\omega$. Thus, the boundary circular plane has a radius of less than 1 unit. From the construction, it is evident for $G \in \prod$,

$$|1 - G_{yr}^*(j\omega)l_m(j\omega)| \geq 1 - |G_{yr}^*(j\omega)|\tilde{l}_m(j\omega), \forall\omega,$$

so that

$$\max_{G \in \prod} |S(j\omega)W(j\omega)| \leq \frac{|(1 - G_{yr}^*(j\omega))W(j\omega)|}{1 - |G_{yr}^*(j\omega)|\tilde{l}_m(\omega)}.$$

It then follows that (2.6) is satisfied if

$$\left|G_{yr}^*(j\omega)\right|\tilde{l}_m(\omega) + \left|(1 - G_{yr}^*(j\omega))W(j\omega)\right| < 1 \quad, \quad \forall\omega,$$

and the proof is completed.

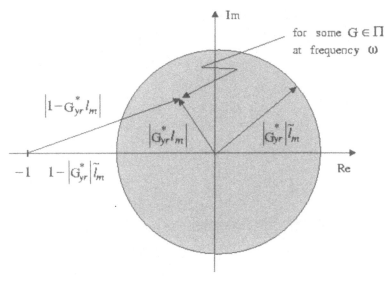

Fig. 2.8. Geometrical construction for derivation of robust performance conditions

CHAPTER 3
GENERALISED PREDICTIVE CONTROLLER

Predictive control is a field which has attracted much research interest and attention over recent decades, judging by the number of publications available, addressing theoretical and application issues. While different control schemes with predictive capability are now available, one of the more common classes of predictive control is probably generalised predictive control (GPC), which was first investigated comprehensively by Clarke *et al.* (1987).

This chapter examines the general form of GPC, illustrates the design of GPC with respect to different cost functions, and provides a stability analysis of the control scheme. It also addresses modification of the basic scheme to adapt to common and practical constraints, such as input saturation, the compensation of disturbance, and strong non-linearity.

3.1 Prediction, Control and Stability

There are three major components in the design of a GPC:

- A model of the system to be controlled. This model is used to predict the system output over the prediction horizon.

- Formulation of the criterion function.

- Minimisation of the criterion function to yield the optimal control output sequence over the prediction horizon.

In the following sub-section, each of the above-mentioned items is discussed in detail.

3.1.1 System Model and Prediction

Consider a system described by the linear state equations:

$$x(t+1) = Ax(t) + Bu(t),$$
$$y(t) \quad = \quad Hx(t), \tag{3.1}$$

where $x(t) \in R^n, y(t) \in R^r$ and $u(t) \in R^m$ are the state, output and control, respectively; A, B, and H are matrices with appropriate dimensions.

The structure of the model (3.1) is used for formulating predictive controllers. First, define a state prediction model of the form:

$$\hat{x}(t+j/t) = A\hat{x}(t+j-1/t) + Bu(t+j-1/t), \; j = 1, 2, ..., p, \tag{3.2}$$

where $\hat{x}(t+j/t)$ denotes the state vector prediction at instant t for instant $t+j$ and $u(\cdot/t)$ denotes the sequence of control vectors within the prediction interval. This model is redefined at each sampling instant t from the actual state vector and the controls previously applied,

$$\hat{x}(t/t) = x(t); \;\; u(t-j/t) = u(t-j); \;\; j = 1, 2, ..., h. \tag{3.3}$$

Applying (3.2) recursively to the initial conditions, the following equations can be obtained:

$$\left. \begin{array}{rl} \hat{y}(t+1/t) = & HAx(t) + HBu(t), \\ \hat{y}(t+2/t) = & HA^2 x(t) + HABu(t) + HBu(t+1), \\ & \vdots \\ \hat{y}(t+p_1/t) = & HA^{p_1} x(t) + HA^{p_1-1} Bu(t) + ... \\ & +HA^{p_1-p_2-1} Bu(t+p_2). \end{array} \right\} \tag{3.4}$$

A composite form of the equation (3.4) can then be written as follows:

$$Y = Gx(t) + F_1 U, \tag{3.5}$$

where

$$Y = [\hat{y}(t+1/t) \; \hat{y}(t+2/t)...\hat{y}(t+p_1/t)]^T,$$

$$G = \begin{bmatrix} HA \\ HA^2 \\ \vdots \\ HA^{p_1} \end{bmatrix},$$

$$F_1 = \begin{bmatrix} HB & 0 & ... & 0 \\ HAB & HB & ... & 0 \\ \vdots & \vdots & ... & \vdots \\ HA^{p_1-1}B & HA^{p_1-2}B & ... & HA^{p_1-p_2-1}B \end{bmatrix},$$

$$U = [u(t), u(t+1), ..., u(t+p_2)]^T. \tag{3.6}$$

3.1.2 Minimisation of the Performance Criterion

The predictive control law is usually formulated to minimise a cost function, also called the performance criterion.

A simple performance criterion that can be used in predictive control design is given by

$$J = \frac{1}{2} \sum_{j=1}^{p_1} [y_d(t+j) - \hat{y}(t+j)]^T Q_j [y_d(t+j) - \hat{y}(t+j)], \tag{3.7}$$

or

$$J = \frac{1}{2}(Y_d - Y)^T Q(Y_d - Y), \tag{3.8}$$

where $y_d(t+j), j = 1, 2, ..., p_1$, is a reference trajectory for the output vector which may be redefined at each sampling instant t, and Q is a non-negative definite matrix.

The solution minimising the performance index may then be obtained by solving

$$\frac{\partial J}{\partial U} = 0, \tag{3.9}$$

from which direct computations may be obtained explicitly as

$$U = (F_1^T Q F_1)^{-1} F_1^T Q [Y_d - G x(t)]. \tag{3.10}$$

However, since F_1 is lower triangular, the prediction horizon p_1 does not affect $u(t)$ and hence the controller obtained when minimising (3.8) is independent of p_1 (Soeterboek, 1992). To consider the effect of the prediction horizon p_1, it has been proven useful to impose a step control sequence together with a cost function such as the one given in (3.8). This type of solution was first introduced in Martin Sanchez (1980) and subsequently analysed in Rodellar (1982).

Under the condition that the control sequence remains constant during the prediction interval, i.e.,

$$u(t) = u(t+1) = ... = u(t+p_1), \tag{3.11}$$

the control action is obtained as:

$$u(t) = (\bar{F}_1^T Q \bar{F}_1)^{-1} \bar{F}_1^T Q [Y_d - Gx(t)], \tag{3.12}$$

where

$$\bar{F}_1 = \begin{bmatrix} HB \\ HAB + HB \\ \vdots \\ H(A^{p_1-1} + A^{p_1-2} + ... + I)B \end{bmatrix}. \tag{3.13}$$

In this case, the prediction horizon p_1 affects $u(t)$ in the control law.

A second way of solving the predictive control problem is to use the following performance criterion:

$$J = \frac{1}{2} \sum_{j=1}^{p_1} [y_d(t+j) - \hat{y}(t+j)]^T Q_j [y_d(t+j) - \hat{y}(t+j)]$$
$$+ \frac{1}{2} \sum_{j=0}^{p_2} u(t+j)^T R_j u(t+j), \tag{3.14}$$

or

$$J = \frac{1}{2}(Y_d - Y)^T Q(Y_d - Y) + U^T RU, \tag{3.15}$$

where $Q \geq 0$ and R is a symmetric matrix.

To obtain the value of the predictive control action $u(t)$, simply substitute (3.5) in (3.15), and minimise the index J to obtain

$$U = (F_1^T Q F_1 + R)^{-1} F_1^T Q [Y_d - Gx(t)]. \tag{3.16}$$

Although (3.16) gives the complete control sequence minimising J over the prediction horizon, only the first value is actually applied to the system as the control signal. Thus, the final control law has the form:

$$u(t) = m_1 [Y_d - Gx(t)], \tag{3.17}$$

with $m_1 = [I_m, 0, 0, ..., 0](F_1^T Q F_1 + R)^{-1} F_1^T Q$ being the first m rows of the matrix $(F_1^T F_1 + R)^{-1} F_1^T$.

A third common way of solving the predictive control problem is to use control increments instead of the control output in (3.15), *i.e.*,

$$J = \frac{1}{2} \sum_{j=1}^{p_1} [y_d(t+j) - \hat{y}(t+j)]^T Q_j [y_d(t+j) - \hat{y}(t+j)]$$

$$+ \frac{1}{2} \sum_{j=0}^{p_2} \Delta u(t+j)^T R_j \Delta u(t+j), \tag{3.18}$$

or

$$J = \frac{1}{2}(Y_d - Y)^T Q(Y_d - Y) + \Delta U^T R \Delta U. \tag{3.19}$$

This performance criterion is used in many predictive controllers. To cope with control increments instead of the control output, the composite equation (3.5) may be rewritten as:

$$Y = Gx(t) + F_{11}\Delta U + F_2 u(t-1), \tag{3.20}$$

where

$$F_{11} = \begin{bmatrix} HB & 0 & \cdots & 0 \\ H(A+I)B & HB & \cdots & 0 \\ \vdots & \vdots & \cdots & \vdots \\ H\Upsilon(p_2)B & H\Upsilon(p_2-1)B & \cdots & HA^{p_1-p_2-1}B \end{bmatrix},$$

$$\Delta U = [\Delta u(t), \Delta u(t+1), ..., \Delta u(t+p_2)]^T,$$

$$F_2 = \begin{bmatrix} HB \\ H(A+I)B \\ \vdots \\ H(A^{p_1-1} + A^{p_1-2} + ... + A^{p_2-p_2-1})B \end{bmatrix},$$

and $\Upsilon(j) = \sum_{k=0}^{k=j} A^{p_1-p_2-1+k}$.

Substituting (3.20) into (3.15) yields:

$$\Delta U = (F_{11}^T Q F_{11} + rI)^{-1} F_{11}^T Q[Y_d - Gx(t) - F_2 u(t-1)]. \tag{3.21}$$

Although (3.21) gives the complete control sequence minimising J over the prediction horizon, only the first m row values are actually applied to the system as the control signal. Thus, the final control law has the form:

$$\Delta u(t) = g_1[Y_d - Gx(t) - F_2u(t-1)], \tag{3.22}$$

with $g_1 = [I_m, 0, 0, ..., 0](F_{11}^T Q F_{11} + R)^{-1}F_{11}^T Q$ being the first m rows of the matrix $(F_{11}^T Q F_{11} + R)^{-1}F_{11}^T Q$.

In general, the performance criterion imposes a compromise between the system output being as close to the reference trajectory as possible and the required control action not being excessive. Using a simpler criterion, the dimension of the control problem can be reduced at the expense of some loss in performances. With a more complex performance criterion, good performances, such as tracking error and robustness, can be achieved, but more parameters must be selected by the designer. In the next section, a stability analysis is given for these predictive control laws.

3.1.3 Stability Analysis

It has been shown in the preceding section that in the design of GPC, a large number of parameters must be selected by the control engineer. In order to make the tuning of GPC easier, a thorough insight into the influence of the design parameters on the stability of the system must be available. In this sub-section, a stability analysis of the predictive control laws descirbed in the preceding section will be provided using a rigorous theoretical analysis.

Lemma 3.1

Consider the nominal system

$$z(t+1) = Az(t), \quad \exists P > 0 \ \forall Q > 0$$

such that

$$A^T P A - P = -Q. \tag{3.23}$$

Then, the perturbed system

$$z(t+1) = Az(t) + \mathbf{m}A_1 z(t) \tag{3.24}$$

is stable if

$$\|\mathbf{m}A_1\| \le \min\left\{\frac{\lambda_{min}(Q)}{2\|A^T P\| + \|P\|}, 1\right\}. \tag{3.25}$$

Proof

Define the Lyapunov function:

$$V(z(t)) = z^T(t)Pz(t). \tag{3.26}$$

The value of $\Delta V(t+1)$ is given by

$$\Delta V(t+1) = z^T(t)(A^T PA - P)z(t) + z^T(t)(2A^T Pm A_1$$
$$+A_1^T m^T Pm A_1)z(t)$$
$$\leq -\lambda_{min}(Q)||z||^2 + (2||A^T P||||m A_1|| + ||m A_1||^2||P||)||z||^2.$$

If $||m A_1|| < 1$, this implies that $||m A_1||^2 \leq ||m A_1||$. Thus, $\Delta V(t+1)$ becomes

$$\Delta V(t+1) \leq -\lambda_{min}(Q)||z||^2 + (2||A^T P|| + ||P||)||m A_1||||z||^2 \tag{3.27}$$

Therefore, if the condition (3.25) holds, the perturbed system (3.24) is stable.

Theorem 3.1

Let the system be described by (3.1) and controlled by the control law in (3.12). Assume the system is open-loop stable. If $Q = I$ (unit matrix), (3.23) holds. R is chosen as $R = rI$. Then, the closed-loop system can be stabilised using (3.12) if

$$||(\bar{F}_1^T \bar{F}_1)^{-1}\bar{F}_1^T G|| < \min\left\{\frac{1}{2||A^T P|| + ||P||}, 1\right\}. \tag{3.28}$$

Proof

It may be proven directly from Lemma 3.1.

Since A is known, P is also known for given $Q = I$. Then, the right side of the condition (3.28) is a constant. For the single-input and single-output (SISO) system, Elshafei *et al.* (1995) has shown that the left side of (3.28) can be set arbitrarily small by increasing the value of p_1. Thus, for SISO system, a minimum value of p_1 can always be found to guarantee the stability by optimising the following objective:

$$\max\{ \| (\bar{F}_1^T \bar{F}_1)^{-1} \bar{F}_1^T G\|\}$$

$$\text{subject to } \|(\bar{F}_1^T \bar{F}_1)^{-1} \bar{F}_1^T G\| < \min\left\{\frac{1}{2\|A^T P\| + \|P\|}, 1\right\}. \quad (3.29)$$

Theorem 3.2.

Let the system be described by (3.1) and controlled by the control law in (3.17). Assume the system is open-loop stable. If $Q = I$, (3.23) holds. R is chosen as $R = rI$. Then, the closed-loop system can be stabilised using (3.17) for a sufficiently large p_1 or r. Furthermore, for a given r, a minimum value of p_1 can be found from the following inequality:

$$\|m_1 G\| < \min\{\frac{1}{2\|A^T P\| + \|P\|}, 1\}. \quad (3.30)$$

Proof

Utilising the result in Lemma 3.1, one can obtain the stability condition (3.30). Since the right side of (3.30) is a positive constant, the inequality (3.30) always hold for a sufficiently large r or p_1. This is because the left side of (3.30) can be set arbitrarily small. Thus, the stability can be guaranteed.

For the control law given by (3.22), letting $Q = I$ and $R = rI$, (3.22) can be rewritten as:

$$u(t) = -g_1 G x(t) + (I_m - g_1 F_2)u(t-1) + g_1 Y_d. \quad (3.31)$$

Note that

$$I_m - g_1 F_2 = (I_m, 0, ..., 0)(F_{11}^T F_{11} + rI)^{-1}(F_{11}^T F_{11} + rI) \begin{pmatrix} I_m \\ 0 \\ \vdots \\ 0 \end{pmatrix} - g_1 F_2$$

$$= (I_m, 0, ..., 0)(F_{11}^T F_{11} + rI)^{-1}[F_{11}^T(F_{11} \begin{pmatrix} I_m \\ 0 \\ \vdots \\ 0 \end{pmatrix} - F_2) + rI]$$

$$= (I_m, 0, ..., 0)(F_{11}^T F_{11} + rI)^{-1} rI, \quad (3.32)$$

since

$$F_{11} \begin{pmatrix} I_m \\ 0 \\ \vdots \\ 0 \end{pmatrix} - F_2 = \begin{pmatrix} HB \\ H(A+I)B \\ \vdots \\ H\Upsilon(p_2)B \end{pmatrix} - F_2$$
$$= 0.$$

Let $\bar{g} = (I_m, 0, ..., 0)(F_{11}^T F_{11} + rI)^{-1}$. The control law (3.31) can be written as

$$u(t) = -g_1 G x(t) + r\bar{g} u(t-1) + g_1 Y_d. \tag{3.33}$$

Combining (3.1) with (3.33) yields

$$\begin{bmatrix} x(t+1) \\ u(t) \end{bmatrix} = \begin{pmatrix} A & B r\bar{g} \\ 0 & \bar{g} r I \end{pmatrix} \begin{bmatrix} x(t) \\ u(t-1) \end{bmatrix} + \begin{pmatrix} -B g_1 G & 0 \\ -g_1 G & 0 \end{pmatrix} \begin{bmatrix} x(t) \\ u(t-1) \end{bmatrix}$$
$$+ \begin{bmatrix} B g_1 Y_d \\ g_1 Y_d \end{bmatrix}. \tag{3.34}$$

Since g_1 and Y_d are bounded, they will not affect the system stability. Thus, the discussions will focus on the following system:

$$\bar{Z}(t+1) = \bar{A}\bar{Z}(t) + \mathbf{g}\bar{Z}(t), \tag{3.35}$$

where

$$\bar{Z}(t) = [x(t), u(t-1)]^T, \quad \bar{A} = \begin{pmatrix} A & B r\bar{g} \\ 0 & \bar{g} r I \end{pmatrix}, \quad \mathbf{g} = \begin{pmatrix} -B g_1 G & 0 \\ -g_1 G & 0 \end{pmatrix}.$$

It follows from the assumption that the system is open-loop stable and $0 \leq |\lambda_i(\bar{g} r I)| < 1$, that the eigenvalues of $\bar{A}$ are inside the unit circle. Hence, the following Lyapunov equation:

$$\bar{A}^T P \bar{A} - P + Q = 0, \tag{3.36}$$

has a solution.

Theorem 3.3

Let the system be described by (3.1) and controlled by the control law in (3.22). Assume the system is open-loop stable. If $Q = I$, (3.36) holds. R is chosen as $R = rI$. Then, the closed-loop system can be stabilised using (3.22) if

$$\|\mathbf{g}\| < \min\left\{\frac{1}{2\|\bar{A}^T P\| + \|P\|}, 1\right\}. \tag{3.37}$$

Proof

From (3.35), the stability of (3.1) with (3.22) follows directly from Lemma 3.1 since $\bar{A}$ is stable.

The theorems, presented in this section can provide insight to stability when there are no constraints. However, nearly every application imposes constraints; actuators are naturally limited in the force they can apply, safety limit states such as temperature, pressure and velocity and efficiency often dictate steady-state operation close to the boundary of the set of permissible states. Predictive control under constraints will be examined in the next section.

3.2 Predictive Control with Input Constraints

The necessity of satisfying input/state constraints is a requirement that frequently arises in control applications. Constraints are dictated, for instance, by physical limitations of the actuators or by the necessity to keep some system variables within safety limits. In predictive control, the constraints can be taken into account explicitly when minimising the predictive control criterion function. This yields the constrained stable predictive control problem. The stability of predictive control with constraints can be investigated in two ways. One is the quadratic programming (QP) method (Garcia and Morshedi, 1984), and the other is the linear matrix inequality (LMI) method (Lu and Arkun, 2000; Casavola *et al.*, 2000). The common drawback of these methods is that on-line optimisation is required. This causes additional computational complexity.

In practice, the most common constraints are on the input and output of the system. These constraints are the so-called hard constraints: no violation of the bounds are allowed. In addition, soft constraints can be considered: violations of the bounds can be allowed temporarily for the satisfaction of other criteria. Saturating actuators are frequently employed as input devices: the valves used in process control systems are typical examples. In this section, predictive control with input saturation will be discussed.

The following linear system with saturating actuators is considered:

$$x(t+1) = Ax(t) + BSat(u(t)) \tag{3.38}$$

and the GPC control in (3.22) is used, *i.e.*,

$$\Delta u(t) = g_1[Y_d - Gx(t) - F_2 u(t-1)], \tag{3.39}$$

where $x \in R^n, u \in R^m$, $\{A, B\}$ is a controllable pair, and g_1 is the first m rows of the matrix $(F_{11}^T Q F_{11} + R)^{-1} F_{11}^T Q$. The saturation function

$$Sat(u) = [sat(u_1), sat(u_2), ..., sat(u_m)]^T$$

is symmetrical with saturating limits $u_i^+, i = 1, 2, ..., m$, and

$$sat(u_i) = \begin{cases} u_i^+, & \text{if} \quad u_i > u_i^+ \\ u_i, & \text{if} -u_i^+ \leq u_i \leq u_i^+ \\ -u_i^+, & \text{if} \quad u_i < -u_i^+ \end{cases}. \tag{3.40}$$

Denote

$$u^+ = [u_1^+, u_2^+, ..., u_m^+]^T. \tag{3.41}$$

Note that since the system (3.38) involves the non-linear function $Sat(\cdot)$, the stability of the closed-loop system (3.38) with (3.39) is dependent on the initial state as well as the characteristics of the non-linear saturation function. The main problem to be addressed here is to derive the conditions on the bounds of the initial state and the characteristics of the saturating non-linearity such that the system (3.38) with (3.39) is stable.

3.2.1 Stability Analysis

This section provides a stability analysis of GPC under input constraints. The following lemmas will be useful in the analysis.

Lemma 3.2

Let $X \in R^{n \times n}$ and $Y \in R^{n \times n}$, and let $y \in R^n$. Then

$$2y^T XYy \leq y^T (\frac{1}{\gamma} XX^T + \gamma Y^T Y)y, \quad \forall \gamma > 0.$$

Lemma 3.3

Consider the following differential equation:

$$x(t+1) = f(x(t)), \; ||x|| \le R^n. \tag{3.42}$$

If there exists a Lyapunov function $V(x) = x^T P x, P > 0$, for which $\Delta V(x) < 0$, and $x \ne 0$, then, for any

$$||x(0)|| < \sqrt{\lambda_{min}(P)/\lambda_{max}(P)}\rho, \tag{3.43}$$

$||x(t)|| \le \rho$ and $\lim_{t \to \infty} ||x(t)|| \to 0$.

Proof

Since $\Delta V(x) < 0$, this implies that

$$V(x) = x^T(t) P x(t) < x^T(0) P x(0).$$

Since $\lambda_{min}(P)||x||^2 \le V(x) \le \lambda_{max}(P)||x||^2$, it follows that

$$||x(t)||^2 \le \frac{\lambda_{max}(P)}{\lambda_{min}(P)} ||x(0)||^2.$$

Thus, the conclusion follows.

To facilitate the analysis, a simple form of conventional GPC with input saturation is first presented.

From (3.39), by denoting the following:

$$\iota_1 = -g_1 Q G, \tag{3.44}$$
$$\iota_2 = g_1, \tag{3.45}$$
$$\iota_3 = I - g_1 F_2, \tag{3.46}$$

(3.39) can be rewritten as

$$u = \iota_1 x(t) + \iota_2 Y_d + \iota_3 u(t-1). \tag{3.47}$$

This equation can also be rewritten as

$$(I - \iota_3) u(t) = \iota_1 x(t) + \iota_2 Y_d - \iota_3 \Delta u(t), \tag{3.48}$$

where $\Delta u(t) = u(t) - u(t-1)$.

Multiplying the inverse of $(I - \iota_3)$ to both sides of (3.48) yields:

$$u(t) = (I - \iota_3)^{-1}\iota_1 x(t) + (I - \iota_3)^{-1}\iota_2 Y_d - (I - \iota_3)^{-1}\iota_3 \Delta u(t). \qquad (3.49)$$

Letting

$$\bar{D}_1 = (I - \iota_3)^{-1}\iota_1,$$
$$\bar{Y}_d = (I - \iota_3)^{-1}\iota_2 Y_d,$$
$$\Delta \bar{u}(t) = -(I - \iota_3)^{-1}\iota_3 \Delta u(t),$$

(3.49) can be rewritten as

$$u(t) = \bar{D}_1 x(t) + \bar{Y}_d + \Delta \bar{u}(t). \qquad (3.50)$$

Assume that $-\Delta \bar{u}^+ \leq \Delta \bar{u}(t) \leq \Delta \bar{u}^+$. The following theorem is given in order to simplify the problem

Theorem 3.4

The problem of the system (3.38) with the saturating control (3.39) can be converted to that of the system:

$$x(t+1) = Ax(t) + BSat(u(t)), \qquad (3.51)$$

with the GPC control

$$u(t) = \bar{D}_1 x(t), \qquad (3.52)$$

where

$$-\mathcal{U} \leq \bar{D}_1 x(t) \leq \mathcal{U}, \qquad (3.53)$$

with

$$\mathcal{U} = min\{u^+ - \Delta \bar{u}^+ + \bar{Y}_d, u^+ - \Delta \bar{u}^+ - \bar{Y}_d\}. \qquad (3.54)$$

Proof

Since saturation control requires that $-u^+ \leq u \leq u^+$, this implies that

$$-u^+ \leq \bar{D}_1 x + \bar{Y}_d + \Delta \bar{u}(t) \leq u^+. \qquad (3.55)$$

This also implies that

$$-u^+ - \bar{Y}_d - \Delta\bar{u}(t) \leq \bar{D}_1 x \leq u^+ -\bar{Y}_d - \Delta\bar{u}(t). \tag{3.56}$$

Using the definition (3.54), the saturation control is converted into

$$-\mathcal{U} \leq \bar{D}_1 x \leq \mathcal{U}. \tag{3.57}$$

In other words, if (3.57) is satisfied, the constraint condition

$$-u^+ \leq u \leq u^+, \tag{3.58}$$

is also met. To illustrate further, it is assumed that, without loss of generality,

$$u^+ - \Delta\bar{u}^+ + \bar{Y}_d \geq u^+ - \Delta\bar{u}^+ - \bar{Y}_d, \tag{3.59}$$
$$\Longrightarrow \quad \bar{Y}_d \geq -\bar{Y}_d. \tag{3.60}$$

This implies that

$$-(u^+ - \Delta\bar{u}^+ - \bar{Y}_d) \leq \bar{D}_1 x \leq (u^+ - \Delta\bar{u}^+ - \bar{Y}_d) \tag{3.61}$$

holds.

Note that $u^+ - \Delta\bar{u}^+ \leq u^+ - \Delta\bar{u}(t)$ since $\Delta\bar{u}(t) \leq \Delta\bar{u}^+$; $-u^+ - \Delta\bar{u}(t) \leq -u^+ + \Delta\bar{u}^+$ since $\Delta\bar{u}(t) \geq -\Delta\bar{u}^+$.

Thus, if (3.57) is satisfied, the following inequality holds:

$$-u^+ - \Delta\bar{u}(t) + \bar{Y}_d \leq \bar{D}_1 x \leq u^+ - \Delta\bar{u}(t) - \bar{Y}_d. \tag{3.62}$$

From (3.60), it follows that

$$-u^+ - \Delta\bar{u}(t) - \bar{Y}_d \leq \bar{D}_1 x \leq u^+ - \Delta\bar{u}(t) - \bar{Y}_d.$$

This implies that

$$-u^+ \leq \bar{D}_1 x + \bar{Y}_d + \Delta\bar{u}(t) \leq u^+.$$

Therefore, the conclusion follows.

The above results show that the saturating GPC problem can be converted equivalently to one involving a linear system subject to input saturation via linear feedback. Next, a novel stabilisation method for linear discrete system subject to input saturation is developed. This method adopts a decomposition of the non-linear function $Sat(\cdot)$. In this case, the controls may be saturated such that the calculated controls may be different from the real control inputs.

Consider the following system:

$$z(t+1) = Az(t) + Bu(t). \tag{3.63}$$

Let the linear feedback control be

$$u(t) = Kz(t), \tag{3.64}$$

where $K = \bar{D}_1$ with $\bar{D}_1$ given in (3.52).

Define

$$\mathcal{U} = [\mathcal{U}_1, \mathcal{U}_2, ..., \mathcal{U}_m]^T, \tag{3.65}$$

and a diagonal matrix

$$\Phi(z) = diag[\phi_1(z), \phi_2(z), ..., \phi_m(z)], \tag{3.66}$$

where

$$\phi_i(z) = \begin{cases} \frac{sat(K_i z)}{K_i z}, & \text{if } K_i z \neq 0 \\ 1, & \text{otherwise} \end{cases} \tag{3.67}$$

with

$$sat(K_i z) = \begin{cases} \mathcal{U}_i & \text{if} \quad u_i > \mathcal{U}_i \\ u_i & \text{if } -\mathcal{U}_i \leq u_i \leq \mathcal{U}_i \\ -\mathcal{U}_i & \text{if} \quad u_i < -\mathcal{U}_i \end{cases}. \tag{3.68}$$

Then, the non-linear function $Sat(\cdot)$ can be rewritten as

$$Sat(u) = -\Phi(z)Kz, \quad u = -Kz. \tag{3.69}$$

Using this, the system (3.63) with the control (3.64) can be written as

$$z(t+1) = A_c z(t) + B(I_m - \Phi(z))Kz, \tag{3.70}$$

where $A_c = A + BK$.

Define a diagonal matrix

$$\bar{\Phi}(\rho) = Diag[\bar{\phi}_1(\rho), \bar{\phi}_2(\rho), ..., \bar{\phi}_m(\rho)], \tag{3.71}$$

where

$$\bar{\phi}_i(\rho) = \begin{cases} \frac{u_i^+}{||K_i||\rho}, & if\ ||K_i||\rho > u_i^+ \\ 1, & otherwise \end{cases}. \tag{3.72}$$

Then, it follows that $0 \le \bar{\phi}_i(\rho) \le \phi_i(z) \le 1$, $i.e.$,

$$\bar{\Phi}(\rho) \le \Phi(z),\ \forall ||z|| < \rho, \tag{3.73}$$

and

$$[I - \Phi(z)]^2 \le [I - \bar{\Phi}(z)]^2,\ \forall z\rho. \tag{3.74}$$

The following theorem provides a sufficient condition to guarantee the stability of the system (3.63) with the control (3.64).

Theorem 3.5

Consider the system (3.63), and let $\bar{\Phi}(z)$ be defined as in (3.66). If there exists a positive matrix P with scalar $\gamma > 0$ such that

$$A_c^T P A_c - P + \gamma A_c{}^T P B B^T P A_c + \frac{1}{\gamma} K^T [I - \bar{\Phi}(\rho)]^2 K$$
$$+ \lambda_{max}(B^T P B) K^T [I - \bar{\Phi}(\rho)]^2 K + \alpha I \le 0, \tag{3.75}$$

then the system (3.63) with the control (3.64) is asymptotically stable for the initial state $||x(0)|| \le \sqrt{\lambda_{min}(P)/\lambda_{max}(P)}\rho$.

Proof

Let $V = z^T P z$. The value of $V(z(t+1)) - V(z(t))$ is

$$V(z(t+1)) - V(z(t)) = z^T [A_c^T + K^T (I - \Phi)^T B^T] P [A_c + B(I - \Phi)K] z$$
$$- z^T P z$$
$$= z^T (A_c^T P A_c - P) z + 2z^T A_c^T P B (I - \Phi) K z$$
$$+ z^T K^T (I - \Phi)^T B^T P B (I - \Phi) K z. \tag{3.76}$$

From Lemma 3.2, it follows that

$$2z^T A_c^T PB(I - \Phi)Kz \leq z^T[\gamma A_c^T PBB^T PA_c + \frac{1}{\gamma}K^T(I - \Phi)^2 K]z$$
$$\leq z^T[\gamma A_c^T PBB^T PA_c + \frac{1}{\gamma}K^T(I - \bar{\Phi})^2 K]z.$$

Substituting the above inequality into (3.76) produces

$$V(z(t+1)) - V(z(t)) \leq z^T[A_c^T PA_c - P + \gamma A_c^T PBB^T PA_c$$
$$+\frac{1}{\gamma}K^T(I - \bar{\Phi})^2 K]z$$
$$+\lambda_{max}(B^T PB)z^T K^T(I - \Phi)^2 Kz$$
$$\leq z^T[A_c^T PA_c - P + \gamma A_c^T PBB^T PA_c$$
$$+\frac{1}{\gamma}K^T(I - \bar{\Phi})^2 K$$
$$+\lambda_{max}(B^T PB)K^T(I - \bar{\Phi})^2 K]z$$
$$\leq -\alpha||z||^2. \tag{3.77}$$

Therefore, from Lemma 3.3, the system (3.63) with the control (3.64) is asymptotically stable for initial state $||x(0)|| \leq \sqrt{\lambda_{min}(P)/\lambda_{max}(P)}\rho$.

Remark 3.1

Theorem 3.5 provides a sufficient condition to guarantee the stability of the constrained GPC. The key point in the control design is the solution of P. The existence of a matrix P has been discussed in Geromel *et al.* (1991).

3.3 Predictive Control with Disturbance Learning

The performance of GPC is inevitably limited and complicated by the presence of disturbances. A model for disturbances is usually assumed to be available in order to adequately deal with them, yet most disturbances are not measurable, although they can be present in a significant way. A linear observer model has been established (Soeterboek, 1992) for a class of deterministic disturbances. Assuming disturbances are known a *priori*, a class of predictive controls has been presented in Hrovat (1991) which includes the extra terms associated with the sequence of disturbances on the prediction interval. By using the Karhunen–Loeve expansion, Rigopoulos *et al.* (1997) developed an approach to identify on-line the dominant disturbance models. Based on neural networks, Draeger *et al.*(1995) presented an extended dynamic matrix predictive control by considering the compensation of disturbances. These methods essentially require an explicit model of the disturbances to be incorporated into the predictive control law.

For a large class of periodic and repeatable disturbances, such as those commonly arising in batch processes and many robotic control applications, GPC can be modified with an appropriate learning scheme, so that the adverse effects of these disturbances can be effectively eliminated with time. Based on this motivation, this section describes the development of a novel framework to combine GPC and a feedforward disturbance learning scheme which is based on the idea of iterative learning control. The basic idea of the control scheme is to improve the control signal for the present operation by feeding back the previous control errors, which are used in the disturbance learning scheme. With repeated cycles, the controller is then able to compensate effectively for the effects of the disturbance, yielding tighter output regulation. The convergence of the learning algorithm can be guaranteed with a sufficiently large number of cycles. An example is given to demonstrate the effectiveness of the algorithm.

3.3.1 GPC Design with Disturbance

The GPC design is based on a linear state formulation described by

$$
\begin{aligned}
x(t+1) &= Ax(t) + Bu(t) + w(t), \\
y(t) &= Cx(t) + v(t),
\end{aligned}
\tag{3.78}
$$

where $x(t), y(t)$ and $u(t)$ are the state, output and control, respectively; $w(t)$ is the input disturbance which is assumed to be repeatable; $v(t)$ is the output uncertainty; A, B, and C are matrices with appropriate dimensions.

First, define a state prediction model of the form:

$$
\hat{x}(t+j/t) = A\hat{x}(t+j-1/t) + Bu(t+j-1/t), \; j = 1, 2, ..., p, \tag{3.79}
$$

where $\hat{x}(t+j/t)$ denotes the state vector prediction at instant t for instant $t + j$ and $u(\cdot/t)$ denotes the sequence of control vectors on the prediction interval. This model is redefined at each sampling instant t from the actual state vector and the control effort previously applied,

$$
\hat{x}(t/t) = x(t); \; u(t-j/t) = u(t-j); \; j = 1, 2, ..., h. \tag{3.80}
$$

Applying (3.79) recursively to the initial conditions, the following equations are obtained:

$$
\left.\begin{aligned}
\hat{y}(t+1/t) &= HAx(t) + HBu(t) + Hw(t), \\
\hat{y}(t+2/t) &= HA^2x(t) + HABu(t) + HBu(t+1) \\
&\quad + HAw(t) + Hw(t+1), \\
&\;\;\vdots \\
\hat{y}(t+p_1 1/t) &= HA^{p_1}x(t) + HA^{p_1-1}Bu(t) + ... \\
&\quad + HA^{p_1-p_2-1}Bu(t+p_2) \\
&\quad + HA^{p_1-1}w(t) + ... + Hw(t+p_1-1).
\end{aligned}\right\}
$$

In Peterson *et al.* (1989), the disturbance vector is assumed to be constant over the prediction horizon. This implies that $w(t) = w(t+1) = ... = w(t + p_1 - 1)$. Following this assumption, a composite form of the above equation can be written as follows:

$$
Y = Gx(t) + F_{11}\Delta U + F_2 u(t-1) + F_3 w(t), \tag{3.81}
$$

where the definitions of Y, G, F_{11}, and F_2 can be found in the equation (3.20), and F_3 is defined as

$$
F_3 = \begin{bmatrix}
H \\
HA + H \\
\vdots \\
HA^{p_1-1} + HA^{p_1-2} + ... + H
\end{bmatrix}.
$$

A linear quadratic performance index is considered and written in the following form:

$$
J = \frac{1}{2}\sum_{j=1}^{p_1}[y_r(t+j) - \hat{y}(t+j)]^T Q_j[y_d(t+j) - \hat{y}(t+j)]
$$

$$
+ \frac{1}{2}\sum_{j=0}^{p_2}\Delta u(t+j)^T R_j \Delta u(t+j), \tag{3.82}
$$

where $p_2 \leq p_1 - 1$, and $y_d(t+j)$ is a reference trajectory for the output vector which may be redefined at each sampling instant t, $\hat{y}(t+j)$ is the prediction of the output trajectory. Q, R are symmetric weighting matrices.

The solution minimising the performance index may then be obtained by solving

$$
\frac{\partial J}{\partial \Delta U} = 0, \tag{3.83}
$$

from which direct computations may be obtained explicitly as

$$\Delta U = (F_1^T Q F_1 + R)^{-1} F_1^T Q [Y_d - Gx(t) - F_2 w(t) - F_3 u(t-1)], \quad (3.84)$$

where $Y_d = [y_d(t+1), y_d(t+2)..., y_d(t+p_1)]^T$.

Although (3.84) provides the complete control sequence minimising J over the prediction horizon, only the first value is actually applied to the system as the control signal as subsequent values will be updated according to prevailing actual states. Thus, the final control law can be written as

$$\Delta u(t) = \mathbf{m}_1 [Y_d - Gx(t) - F_2 w(t) - F_3 u(t-1)], \quad (3.85)$$

$\mathbf{m}_1$ being the first row of the matrix $(\bar{F}_1^T Q \bar{F}_1 + R)^{-1} \bar{F}_1^T Q$.

Since the performance of predictive control depends also on how effectively the disturbances present can be accounted for, it is useful to be able to obtain an estimate of $w(t)$. With the assumption that the occurrence of the disturbance is periodic and repeatable, and assuming further that a feedforward estimate of the disturbance is available, an improved GPC control law is given by

$$u(t) = \mathbf{m}_1 [Y_d - Gx(t) - F_2 \hat{w}_i^f(t)] + (I - \mathbf{m}_1 F_3) u(t-1). \quad (3.86)$$

By using the following notation:

$$D_0 = -\mathbf{m}_1 F_2, \quad (3.87)$$
$$D_1 = -\mathbf{m}_1 QG, \quad (3.88)$$
$$D_2 = \mathbf{m}_1, \quad (3.89)$$
$$D_3 = I - \mathbf{m}_1 F_3, \quad (3.90)$$

(3.86) can be rewritten as

$$u(t) = D_0 \hat{w}_i^f(t) + D_1 x(t) + D_2 Y_d + D_3 u(t-1). \quad (3.91)$$

Remark 3.2

The various predictive control schemes, such as Garcia and Morari (1982), Mosca and Zhang (1992), and Huang $et\ al.$ (1999a,b,c), and their analysis remain applicable to (3.91). Note that the main difference in the present algorithm is the inclusion of an additional feedforward term $D_0 \hat{w}_i^f(t)$. The stability properties of the enhanced algorithm are identical to the conventional predictive control schemes.

3.3.2 Disturbance Learning Scheme

The main purpose of $\hat{w}_i^f(t)$ is to provide an estimate for $w(t)$. The iterative learning control (ILC) is used to achieve this. The ILC method was first proposed by Arimoto $et\ al.$ (1984). To date, many of the learning algorithms developed have been used in applications on robot control where the system is required to execute the same task repetitively (see Draeger $et\ al.$, 1995; Rigopoulos $et\ al.$, 1997). In this section, a new learning algorithm combined with GPC is presented. The following two assumptions are necessary in the following developments.

Assumption 3.1

The disturbance $w(t)$ is repeatable on $[0, N]$, and each element $w(t)$ is bounded.

Assumption 3.2

The state is assumed to be re-initialised exactly at $x_d(0)$, $i.e.$, $x_i(0) = x_d(0)$, where $x_d(0)$ is the desired initial state.

The feedforward learning algorithm is given by:

$$\hat{w}_{i+1}^f(t) = \hat{w}_i^f(t) + L(t)e_i(t+1), \tag{3.92}$$

where $L(t)$ is the learning gain, and $e_i(t) = y_d(t) - y_i(t)$. Clearly, for the ith learning cycle, the input $\hat{w}_{i+1}^f$ will use its previous estimate plus the error information at time $t+1$ during the ith iteration.

Consider the following closed-loop system obtained after substituting the predictive control law (3.91):

$$x(t+1) = (A + BD_1)x(t) + BD_0\hat{w}_i^f(t) + w(t)$$
$$+BD_2Y_d + BD_3u(t-1), \tag{3.93}$$
$$y(t) = Hx(t). \tag{3.94}$$

For convergence analysis, it is necessary to combine the control and state equations to produce a composite form:

$$\begin{bmatrix} u(t) \\ x(t+1) \end{bmatrix} = \begin{bmatrix} D_3 & D_1 \\ BD_3 & A+BD_1 \end{bmatrix} \begin{bmatrix} u(t-1) \\ x(t) \end{bmatrix} + \begin{bmatrix} D_0 \\ BD_0 \end{bmatrix} \hat{w}_i^f(t)$$
$$+ \begin{bmatrix} 0 \\ I \end{bmatrix} w(t) + \begin{bmatrix} D_2 \\ BD_2 \end{bmatrix} Y_d. \tag{3.95}$$

Let

$$\bar{x} = [u(t), x(t+1)]^T, \tag{3.96}$$

$$\bar{A} = \begin{bmatrix} D_3 & D_1 \\ BD_3 & A + BD_1 \end{bmatrix}, \tag{3.97}$$

$$\bar{B}_1 = \begin{bmatrix} D_0 \\ BD_0 \end{bmatrix}, \tag{3.98}$$

$$\bar{B}_2 = \begin{bmatrix} 0 \\ I \end{bmatrix}, \tag{3.99}$$

$$\bar{B}_3 = \begin{bmatrix} D_2 \\ BD_2 \end{bmatrix}, \tag{3.100}$$

$$\bar{H} = [0, H]. \tag{3.101}$$

(3.95) can be rewritten as

$$\bar{x}_i(t+1) = \bar{A}\bar{x}_i(t) + \bar{B}_1 \hat{w}_i^f(t) + \bar{B}_2 w(t) + \bar{B}_3 Y_d, \tag{3.102}$$

$$y = \bar{H}\bar{x}, \tag{3.103}$$

where i denotes the ith repetitive operation of the system and $t \in [0, N]$ is the time interval within a period of the periodic disturbance .

Lemma 3.4 (Chen, 1999; Theorem 5.D4)

The time-invariant dynamic linear system $z(k+1) = \Phi z(k)$ is asymptotically stable if and only if all eigenvalues of Φ have magnitudes less than 1, i.e., $|\lambda_j(\Phi)| < 1$.

Theorem 3.6

For the system (3.93),(3.94), with Assumptions 3.1–3.2, by using the feedforward learning control law (3.92), $\hat{w}_i^f(t)$ will converge to the unknown periodic disturbance $w(t)$ on $[0, N]$ as $i \to \infty$, if $|\lambda_i[I - L(t)\bar{H}\bar{B}_1]| < 1$ for all $t \in [0, N]$.

Proof

By defining the tracking error $\delta w_i^f(t) = w(t) - \hat{w}_i^f(t)$, it follows that

$$\delta w_{i+1}^f(t) = \delta w_i^f(t) - L(t)e_i(t+1). \tag{3.104}$$

To obtain $e_i(t+1)$, define $\delta \bar{x}_i(t) = \bar{x}_d(t) - \bar{x}_i(t)$. For the desired reference trajectory $\bar{x}_d(t)$, it is shown that $\hat{w}_i^f(t) = w(t)$. Thus,

$$\bar{x}_d(t+1) = \bar{A}\bar{x}_d(t) + \bar{B}_1 w(t) + \bar{B}_2 w(t) + \bar{B}_3 Y_d. \tag{3.105}$$

With (3.93) and (3.105), it follows that

$$\delta\bar{x}_i(t+1) = \bar{A}\delta\bar{x}_i(k) + \bar{B}_1 \delta w_i^f(t). \tag{3.106}$$

$e_i(t+1)$ can be computed from the following equation:

$$e_i(t+1) = y_d(t+1) - y_i(t+1) = \bar{H}\delta\bar{x}_i(t+1). \tag{3.107}$$

Substituting (3.107) with (3.106) into (3.104) yields

$$\begin{aligned}
\delta w_{i+1}(t) &= \delta w_i(t) - L(t)\bar{H}[\bar{A}\delta\bar{x}_i(t) + \bar{B}_1 \delta w_i(t)] \\
&= (I - L\bar{H}\bar{B}_1)\delta w_i(t) - L(t)\bar{H}\bar{A}\delta\bar{x}_i(t).
\end{aligned} \tag{3.108}$$

From (3.106), one obtains

$$\begin{aligned}
\delta\bar{x}_i(t) &= \bar{A}^t \delta\bar{x}_i(0) + \bar{A}^{t-1}\bar{B}_1 \delta w_i(0) + \bar{A}^{t-2}\bar{B}_1 \delta w_i(1) + ... \\
&\quad + \bar{B}_1 \delta w_i(t-1).
\end{aligned} \tag{3.109}$$

By assuming $\delta\bar{x}_i(0) = 0$, (3.109) may be expanded as

$$\begin{aligned}
\delta w_{i+1}(0) &= (I - L(0)\bar{H}\bar{B}_1)\delta w_i(0) \\
\delta w_{i+1}(1) &= (I - L(1)\bar{H}\bar{B}_1)\delta w_i(1) - L(1)\bar{H}\bar{A}\bar{B}_1 \delta w_i(0) \\
&\quad \vdots \\
\delta w_{i+1}(N) &= (I - L(N)\bar{H}\bar{B}_1)\delta w_i(N) - L(N)\bar{H}\bar{A}[\bar{A}^{N-1}\bar{B}_1 \delta w_i(0) \\
&\quad + \bar{A}^{N-2}\bar{B}_1 \delta w_i(1) + ... + \bar{B}_1 B \delta w_i(N-1)].
\end{aligned} \tag{3.110}$$

For simplicity, the following notation is used:

$$\begin{aligned}
\psi_{j,j} &= I - L(j)\bar{H}\bar{B}_1, j = 0, 1, ...N, \\
\psi_{1,0} &= -L(1)\bar{H}\bar{A}\bar{B}_1, \\
&\quad \vdots \\
\psi_{N,0} &= -L(N)\bar{H}\bar{A}\bar{A}^{N-1}\bar{B}_1, \\
\psi_{N,1} &= -L(N)\bar{H}\bar{A}\bar{A}^{N-2}\bar{B}_1, \\
&\quad \vdots \\
\psi_{N,N-1} &= -L(N)\bar{H}\bar{A}\bar{B}_1.
\end{aligned}$$

The above equations can be rewritten in composite form:

$$\mathcal{W}_{i+1} = \Psi \mathcal{W}_i, \tag{3.111}$$

where

$$\mathcal{W}_i = [\delta w_i(0), \delta w_i(1), ..., \delta w_i(N)]^T, \tag{3.112}$$

$$\Psi = \begin{bmatrix} \psi_{0,0} & 0 & .... & 0 \\ \psi_{1,0} & \psi_{1,1} & ... & 0 \\ \vdots & \vdots & \vdots & \vdots \\ \psi_{N,0} & \psi_{N,1} & ... & \psi_{N,N} \end{bmatrix}. \tag{3.113}$$

It should be emphasised that the matrix Ψ is an $Nm \times Nm$ constant matrix with respect to the iteration i. Thus, the system becomes a discrete time-invariant system. According to Lemma 3.4, $\mathcal{W}_i$ is convergent if Ψ is stable, i.e., all $|\lambda_k[\Psi]| < 1$.

Since the matrix Ψ is a lower block triangular one, it follows that

$$\lambda_k[\Psi] = \cup_{j=0}^N \{\lambda_k[\psi_{j,j}]\}. \tag{3.114}$$

This implies that $\mathcal{W}_i$ is convergent if $\psi_{j,j}$ for each $j = 0, 1, 2, ...N$ is a stability matrix, i.e., if all of its eigenvalues are inside the unit circle. The proof is completed.

Theorem 3.7

For the system (3.93),(3.94), with Assumptions 3.1–3.2, by using the feed-forward learning control law (3.92), there exists a sequence of learning gains $L(t), t = 0, 1, 2, ..., N$, such that $\hat{w}_i^f(t)$ converges to the unknown periodic disturbance $w(t)$ on $[0, N]$ as $i \to \infty$, if $\bar{H}\bar{B}_1$ is full column rank.

Proof

Theorem 3.6 shows that the algorithm is convergent if the matrices $\lambda_i[I - L(t)\bar{H}\bar{B}_1]$ are stable. For each t, the matrix $L(t)$ which stabilises the corresponding matrix $I - L(t)\bar{H}\bar{B}_1$, exists if pair $\bar{H}\bar{B}_1$ has full column rank. Thus, the conclusion follows.

Remark 3.3

The present method differs from those previously published (Soeterboek, 1992; Hrovat, 1991; Rigopoulos et al., 1997). As can be seen, the algorithm does not need an explicit disturbance model, irregardless of whether it is a linear or non-linear one. It requires only the tracking experience from previous cycles.

3.3.3 Simulation Example

The control scheme is applied to the velocity control of a DC servomotor. The mechanical and electrical dynamics of a standard DC motor can be expressed as follows (Shiro *et al.*, 1995; Fujimoto and Kawamura, 1995):

$$M\ddot{x} + D\dot{x} + T_l = T_m, \tag{3.115}$$

$$K_E\dot{x} + L_a\frac{dI_a}{dt} + R_aI_a = u, \tag{3.116}$$

$$T_m = K_TI_a, \tag{3.117}$$

where x denotes position; M, D, T_m, T_l denote the mechanical parameters: slide mass, viscosity constant, generated force and load force respectively; u, I_a, R_a, L_a denote the electrical parameters: input DC voltage, armature current, armature resistance and armature inductance respectively; K_T denotes the electrical–mechanical energy conversion constant.

Since the electrical time constant is small (compared to the mechanical time constant), the electrical transients decay very rapidly and $L_a\frac{dI_a}{dt} \sim 0$ (see Chapter 5, Nasar and Boldea, 1987; Fujimoto and Kawamura, 1995). Thus, the following simplified equation is obtained:

$$\ddot{x}(t) = -\frac{K_1}{M}\dot{x}(t) + \frac{K_2}{M}u(t) - \frac{1}{M}T_l, \tag{3.118}$$

where

$$K_1 = \frac{K_EK_T + R_aD}{R_a}, \quad K_2 = \frac{K_T}{R_a}. \tag{3.119}$$

Extraneous non-linear effects which may be present in the servo system are also included in T_l. However, T_l is assumed to be bounded

Consider the parameters of an industrial grade MT22G2 DC servomotor presented in Table 3.1. For the simulation, it is assumed that the system parameters D, M are 0.712N/m/s and 0.59kg, respectively, since they are unknown. The sampling time is 0.005s. The control objective is to control the output close to the desired velocity trajectory. The desired position trajectory is shown in Fig. 3.1 (solid line).

Table 3.1. Parameters of MT22G2 DC servomotor

Contents	Units	$LS1 - 24$
Force constant (K_T)	N/Amp	0.10
Resistance (R_a)	Ohms	0.63
Back EMF(K_E)	volt/m/s	0.10
Viscosity constant (D)	–	–
Length of travel	m	609.6
Slide weight	–	–
Armature inductance(L_a)	mh	2.1
Peak current(I_p)	Amp	42
Typical velocity	RPM	5000

First, to illustrate the performance of the controller, simulation results without disturbance affecting the system are presented. This scheme may be referred to as the conventional GPC scheme. In this case, choose the following design parameters: $p_1 = p_2 = 6$, $Q = I$, and $R = 10^{-4}I$, where I is the 6×6 unit matrix. The dotted line in Fig. 3.1 (top) shows the tracking performance. It is observed that the tracking performance is satisfactory. Next, a periodic disturbance described by $w(t) = 3\sin(t)$ is introduced to the system. For comparison, pure GPC without disturbance learning is first presented. The dotted line in Fig. 3.2 shows the output tracking. It is observed that the tracking performance has been significantly degraded by the presence of the disturbance.

For the learning-enhanced GPC algorithm, the learning gains are designed to meet the conditions of Theorem 3.6. The simulation results are shown in Fig. 3.3. It can be seen that the velocity tracking (dotted line in the top of Fig. 3.3) is improved significantly. The reason for improvement is that the feedforward learning can adequately estimate the unknown disturbance in a non-parametric form. The learning capability is also shown in Fig. 3.3 (middle), where the dotted line is the estimate of $w(t)$ and the solid line is the actual disturbance. When the learning gains are increased but still meet the conditions of Theorem 3.6, the tracking performance is further improved as shown in Fig. 3.4. This further verifies the conclusions of Theorem 3.6.

3.4 Non-linear Predictive Control

Most predictive controls are based on linear systems. For non-linear systems, the controller requires continuous on-line model identification to take care of system non-linearity. The identification of an on-line model sometimes requires injection of additional disturbance signals into the system for sufficiently rich excitation of system dynamics. This may not be acceptable in

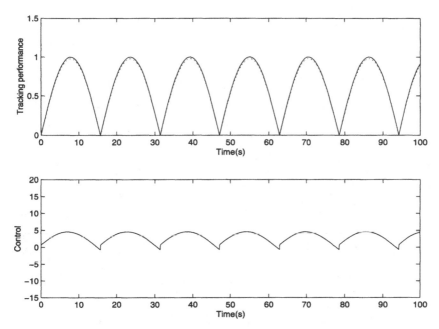

Fig. 3.1. Control performance of GPC (no disturbance present): solid line is desired trajectory and dotted line is actual response

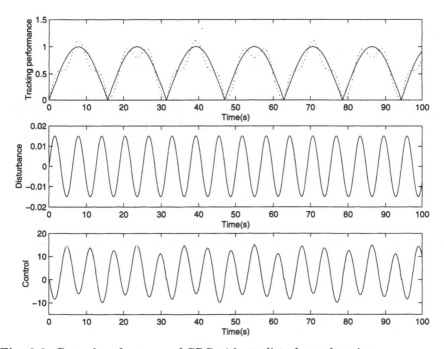

Fig. 3.2. Control performance of GPC without disturbance learning

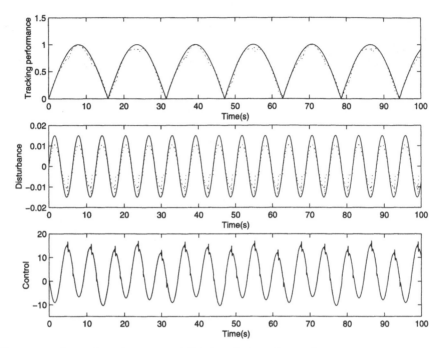

Fig. 3.3. Control performance of GPC with disturbance learning

certain cases for reasons of reliability and safety. On the other hand, the consequences of failure of the parameter estimation algorithm are severe in that the estimator can act to accentuate system disturbances.

Ideally, a non-linear system requires a non-linear model predictive controller for optimum performance (Hernandez and Arkun, 1990; Fisher *et al.*, 1998). Real-time dynamic optimisation using a non-linear model and non-linear optimisation technique is a complex procedure and computationally highly taxing. Therefore, in this section, an approach which aims to reduce the computational burden is presented. Basically, linear models are extracted from the non-linear models and provided to the linear predictive controller. In order to take into account system non-linearity, a neural-network-based technique is used to develop a non-linear dynamic model from empirical data. This model is used to create a linear dynamic model at each time instant. The time-variant linear model obtained is utilised in GPC.

Figure 3.5 shows the concept of predictive learning control based on dynamic linearisation of a neural network model. At the lower control level, a GPC is employed. The parameters of the time-variant linear model used for prediction are updated by dynamic linearisation of the neural network. The approach is similar to parameter-adaptive control. Now, the complete non-linear model information is contained in the neural network model and

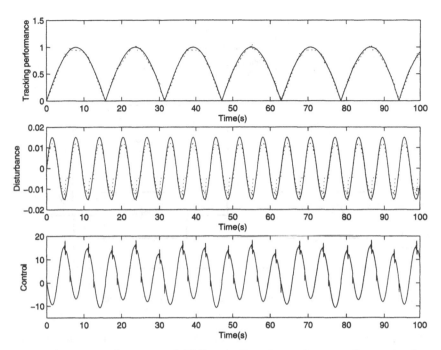

Fig. 3.4. Control performance of GPC with disturbance learning (increasing learning gains)

an on-line adaptation of the linear model is superfluous. Hence, there is no danger arising from possible failure of an estimation algorithm.

3.4.1 System Model

The non-linear dynamic systems can be described in the discrete time domain by

$$y(t+1) = f(y(t), ..., y(t-n_y), u(t), ..., u(t-n_u)), \qquad (3.120)$$

where t is the time axis, and $y(t)$ and $u(t)$ are the system output and input, respectively. The unknown function $f(\cdot)$ has to be approximated from measured data. In this section, a neural network model is applied to this task.

Neural networks (NNs) are general tools for modeling non-linear functions since they can approximate any non-linear function to any desired level of accuracy (Narendra and Parthasarathy, 1990). The potential of NNs for practical applications lies in the following properties they possess: (1) they can be used to approximate any continuous mapping, (2) they achieve this approxi-

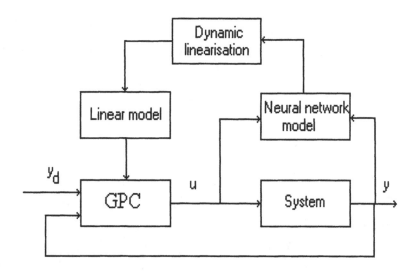

Fig. 3.5. Predictive control scheme based on non-linear models

mation through learning, (3)parallel processing and non-linear interpolation can easily be accomplished with NNs.

One of the well-developed NNs is a three-layer perceptron NN. Figure 3.6 shows the basic structure of this network. It has been proven that a three-layer NN can be used to approximate any continuous function to the desired accuracy. Given $X \in R^N$, a three-layer NN has a net output given by

$$
X_{2k} = f(X; \mathcal{W}) = \sum_{h=1}^{N_1} \sigma \left[\sum_{j=1}^{N} \left[W_{1jk}\sigma \left[\sum_{i=1}^{N_0} W_{ij}X_i + \theta_{0j} \right] + \theta_{1k} \right] \right] + \theta_2,
$$
$$
k = 1, ...N_2, \tag{3.121}
$$

where χ is a regression vector given by

$$
\chi = [y(t), ..., y(t - n_y), u(t), ..., u(t - n_u)]^T, \tag{3.122}
$$

$\sigma(\cdot)$ being the activation function, W_{ij} the first-to-second layer interconnection weights, and W_{1jk} the second-to-third layer interconnection weights. θ_{1j}, θ_{2k}, are threshold offsets. It is usually desirable to adapt the weights and thresholds of the NN off-line or on-line in real-time to achieve the required approximation performance of the net. That is, the NN should exhibit a "learning behaviour". The back-propagation (BP) algorithm has been used in designing the NN model. The BP algorithm is based on the gradient algorithm to minimise the network-output error, and is derived from the special structure of the networks. Using the structure in Fig. 3.6, computation of the NN's output and updating of the NN's weights are summarized as follows:

A. Compute the output of the HIDDEN layer, X_{1j}

$$\bar{X}_{1j} = \frac{1}{1 + \exp(-\bar{O}_{1j} + \bar{\theta}_{1j})},$$

where $\bar{O}_{1j} = \sum_{i=1}^{N} W_{ij} X_i$, and X_i is the input (or input sample) of the NN.

B. Compute the output of OUTPUT layer, X_{21}

$$X_{21} = \frac{1}{1 + \exp(-O_{21} - \theta_{21})}, \tag{3.123}$$

where X_{21} is the output of the NN, and $O_{21} = \sum_{j=1}^{N_1} W_{1j1} X_{1j}$.

C. Update the weights from HIDDEN to OUTPUT layer, W_{1j1} according to

$$W_{1j1}^{t+1} = W_{1j1}^{t} + \eta_1 \delta_{11} X_{1j}, \tag{3.124}$$

where $\delta_{11} = -(X_{21}^{d} - X_{21})$ with X_{21}^{d} being the desired output (or output sample), and X_{21} being the NN output.

D. Update the weights from INPUT to HIDDEN layer, W_{ij}

$$W_{ij}^{t+1} = W_{ij}^{t} + \eta_2 \delta_j \bar{X}_i, \tag{3.125}$$

where $\delta_j = [\delta_{11} W_{1j1}] X_{1j} (1 - X_{1j})$.

E. Update the thresholds, θ_{21}, θ_{1j}

$$\theta_{21}^{t+1} = \theta_{21}^{t} + \eta_{1\theta} \delta_{11}, \theta_{1j}^{t+1} = \theta_{1j}^{t} + \eta_{2\theta} \delta_j, \tag{3.126}$$

where $\eta_{1\theta}$ and $\eta_{2\theta}$ are gain factors.

A terminating condition for the training process is usually formulated, that is

$$\mathcal{E} = \frac{1}{2} \sum_{l=1}^{M} (X_{21}^{dl} - X_{21}^{l})^2, \tag{3.127}$$

where l represents the sample number. The iterative weights tuning process is terminated when the error converges to within a specified threshold. Thus, the optimum weights W can be obtained. The training is usually a tradeoff in terms of the quality of fit and the iteration time. Since the training process can be done off-line, more emphasis may be given to deriving a better fit at the expense of incurring a longer tuning time.

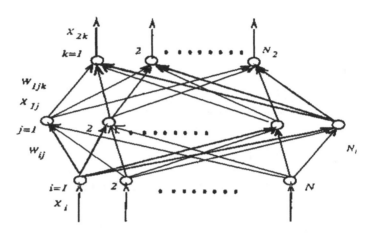

Fig. 3.6. Basic structure of a multilayer NN

3.4.2 Dynamic Linearisation

Based on the NN model, linear models are extracted at each time instant. In contrast with classical linearisation which is only performed in equilibria along the static input output mapping, the system is linearised at each time instant regardless of whether the system is in a steady or in a transient state. Hence, actual information about the system dynamics is available in all states of operation.

Now, the objective is to calculate the time-invariant parameters $a_i(t)$ and $b_i(t)$ of the linear transfer function

$$G(z,t) = \frac{b_1 z^{-1} + b_2 z^{-2} + ... + b_n z^{-n}}{1 + a_1 z^{-1} + a_2 z^{-2} + ... + a_n z^{-n}} \qquad (3.128)$$

for arbitrary system states.

The parameters are obtained using a first-order Taylor series approximation of the non-linear model:

$$a_k(t) = -\frac{\partial y(t)}{\partial y(t-i)}|_{x=x(t)}, \tag{3.129}$$

$$b_k(t) = -\frac{\partial y(t)}{\partial u(t-i)}|_{x=x(t)}. \tag{3.130}$$

The equations (3.129)–(3.130) can be computed from the measurement of the system output. In other words, a function of the network output is known since the detailed structure and parameters of the mapping $f(\cdot)$ are known. Using the gradient algorithm, the first-order Taylor series approximation can be obtained.

Based on the structure of the above three-layer NN, the derivatives of (3.129), (3.130) can be deduced by

$$\frac{\partial X_{21}}{\partial O_{21}} = X_{21}(1 - X_{21}), \tag{3.131}$$

$$\frac{\partial O_{21}}{\partial X_{1j}} = W_{1j1}, \tag{3.132}$$

$$\frac{\partial X_{1j}}{\partial O_{1j}} = X_{1j}(1 - X_{1j}), \tag{3.133}$$

$$\frac{\partial O_{1j}}{\partial X_i} = W_{ij}. \tag{3.134}$$

Applying the chain rule of the gradient algorithm yields

$$\frac{\partial O_{21}}{\partial X_i} = \sum_{j=1}^{N_1} \frac{\partial O_{21}}{\partial X_{1j}} \frac{\partial X_{1j}}{\partial O_{1j}} \frac{\partial O_{1j}}{\partial X_i}$$

$$= \sum_{j=1}^{N_1} X_{21}(1 - X_{21})W_{1j1}X_{1j}(1 - X_{1j})W_{ij} \tag{3.135}$$

Once the remaining partial derivatives in (3.129) and (3.130) are determined, the equations yield the time-variant parameters depending on $x(t)$.

The model (3.128) can also be written in the form

$$x(t+1) = Ax(t) + Bu(t), \tag{3.136}$$

$$y(t) = Hx(t), \tag{3.137}$$

where

$$A = \begin{bmatrix} -a_1 & 1 & ... & 0 \\ -a_2 & 0 & ... & 0 \\ \vdots & \vdots & ... & \vdots \\ -a_{n-1} & 0 & ... & 1 \\ -a_n & 0 & ... & 0 \end{bmatrix},$$ (3.138)

$$B = \begin{bmatrix} b_1 \\ b_2 \\ \vdots \\ b_{n-1} \\ b_n \end{bmatrix},$$ (3.139)

$$H = [1\ 0\ ...0].$$ (3.140)

3.4.3 Predictive Control Scheme

The prediction of the system output from 1 to p_1 can be written in vector representation as given in (3.20):

$$Y = Gx(t) + F_{11}\Delta U + F_2 u(t-1).$$ (3.141)

Consider the following finite horizon objective function:

$$J = (Y_d - Y)^T(Y_d - Y) + r\Delta U_i^T \Delta U_i$$ (3.142)

where $Y_d = [y_d(t+1/t), y_d(t+2/t), ..., y_d(t+p_1/t)]^T$.

At every time t GPC control is

$$\Delta U = (F_{11}^T F_{11} + rI)^{-1} F_{11}^T [Y_d - Gx(t) - F_2 u(t-1)].$$ (3.143)

The first m row values are actually applied to the system. The general optimisation problem subject to constraints can be posed as

$$min\ J = (Y_d - Y)^T(Y_d - Y) + r\Delta U^T \Delta U$$ (3.144)

subject to

$$Y_{min} \le Y \le Y_{max},$$
$$U^{low} \le U \le U^{hi},$$
$$\Delta U^{low} \le U \le \Delta U^{hi}.$$

This problem can be solved using a quadratic optimisation technique. To this end, the objective function and constraints are required to be defined as functions of manipulating variables.

Using (3.143) and after removing the constant terms, the complete optimisation problem can be written in the following quadratic form:

$$min\ J = (Y_d - Y)^T (Y_d - Y) + r\Delta U^T \Delta U \tag{3.145}$$

subject to

$$\begin{bmatrix} F_1 \\ -F_1 \\ I \\ -I \\ I \\ -I \end{bmatrix} \Delta U \leq \begin{bmatrix} Y_{max} - Gx(t) - F_2 u(t-1) \\ -Y_{min} + Gx(t) + F_2 u(t-1) \\ \Delta U^{hi} \\ -\Delta U^{low} \\ U^{hi} - u(t-1) \\ -U^{low} + u(t-1) \end{bmatrix} \tag{3.146}$$

Together with the execution time of the local dynamic linearisation, the computational effort is significantly lower than for predictive control directly based on a non-linear model. The QP algorithm must be applied to solve the optimisation problem (Garcia and Morshedi, 1984).

CHAPTER 4

OBSERVER-AUGMENTED GENERALISED PREDICTIVE CONTROLLER

One of the main drawbacks of the generalised predictive control (GPC) is the requirement that the states of the system are available. Since this requirement may be quite restrictive in applications, it is desirable to be able to derive, at least, estimates of the states using an observer (Ricker, 1990; Yoon and Clarke, 1995).

In this chapter, a generalised predictive observer-controller (GPOC), combining the predictive control structure and the state observer, is developed for a general class of systems with input delay and non-measurable states. These include systems which are subject to noisy and non-measurable disturbances since the states of such systems are essentially unknown even if they are measurable. Based on a state-space model with input time delay and under the influence of system disturbances, a GPC solution is first formulated in detail. For the purpose of states prediction under noisy conditions, a steady-state Kalman filter is designed as the states observer. Such a hybrid control structure retains the benefits of a GPC while expanding the domain of practical applications of the controller. From a practical application viewpoint, it is also interesting to examine the closed-loop properties from perspectives of stability and robustness of the designed GPOC. The chapter shows that with appropriate selection of the prediction horizon and weighting matrices, the stability of the closed loop can be ensured even in the face of small perturbations due to modelling errors. Finally, the GPOC is applied to an injection moulding situation to address the problems of input delay and disturbance noise.

4.1 Predictive Control Solution for Time-delay Systems

In this section, a predictive control solution based on a GPC incorporating a Kalman filter states observer is presented for a general single-input/single-output (SISO) linear system with time delay and non-measurable states.

4.1.1 System Model with Time-delay and Non-measurable Noise

The system under consideration can be represented as the following model:

$$x(t+1) = Ax(t) + Bu(t-h) + D\xi(t), \tag{4.1}$$
$$y(t) = Hx(t) + \xi(t), \tag{4.2}$$

where h is a pure time delay related to the system input, and ξ is a zero mean white noise sequence with covariance matrix ζ.

A controlled autoregressive integrated moving average (CARIMA) model is often used in practical control problems, $i.e.$,

$$a(q^{-1})y(t) = b(q^{-1})u(t-h) + c(q^{-1})\xi,$$
$$a(q^{-1}) = 1 + a_1 q^{-1} + ... + a_n q^{-n}, a_n \neq 0,$$
$$b(q^{-1}) = b_1 q^{-1} + b_2 q^{-2} + ... + b_n q^{-n},$$
$$c(q^{-1}) = 1 + c_1 q^{-1} + ... + c_n q^{-n},$$

where q^{-1} is the unit delay operator, h is the pure time delay related to the system input, and ξ is a zero mean white noise sequence. The CARIMA model can also be written in the above state space form (4.1),(4.2), where

$$A = \begin{bmatrix} -a_1 & 1 ... 0 \\ -a_2 & 0 ... 0 \\ \vdots & \vdots ... \vdots \\ -a_{n-1} & 0 ... 1 \\ -a_n & 0 ... 0 \end{bmatrix}, B = \begin{bmatrix} b_1 \\ b_2 \\ \vdots \\ b_{n-1} \\ b_n \end{bmatrix}, D = \begin{bmatrix} c_1 - a_1 \\ c_2 - a_2 \\ \vdots \\ c_{n-1} - a_{n-1} \\ c_n - a_n \end{bmatrix},$$

$$H = [1\ 0\ ...0].$$

In this chapter, the main concern is with the model (4.1),4.2 for predictive control design and analysis.

4.1.2 Prediction with Input Delay and Noise

The structure of the model (4.1),(4.2) is useful in the formulation of predictive controllers. First define a state prediction model of the form:

$$\hat{x}(t+j/t) = A\hat{x}(t+j-1/t) + Bu(t+j-1-h/t), \ \ j = 1, 2, ..., p+h+1, \tag{4.3}$$

where $\hat{x}(t+j/t)$ denotes the state vector prediction at instant t for instant $t+j$ and $u(\cdot/t)$ denotes the sequence of control vectors on the prediction interval. This model is redefined at each sampling instant t from the actual state vector and the controls previously applied, *i.e.*,

$$\hat{x}(t/t) = x(t); \quad u(t-j/t) = u(t-j); \quad j = 1, 2, ..., h. \tag{4.4}$$

A linear quadratic performance index written in the following form is considered:

$$J = \frac{1}{2} \sum_{j=1}^{h+p+1} [y_d(t+j) - \hat{y}(t+j)]^T Q_j [y_d(t+j) - \hat{y}(t+j)]$$

$$+ \frac{1}{2} \sum_{j=0}^{p} \Delta u(t+j)^T R_j \Delta u(t+j), \tag{4.5}$$

where $y_d(t+j)$ is a reference trajectory for the output vector which may be redefined at each sampling instant t, $\hat{y}(t+j)$ is a prediction output trajectory and $\Delta u(t+j) = u(t+j) - u(t+j-1)$. Q, R are symmetric weighting matrices and p denotes the prediction horizon.

Applying (4.3) recursively to the initial conditions, it follows that

$$\left.\begin{aligned}
\hat{y}(t+1/t) &= H\hat{x}(t+1/t), \\
\hat{y}(t+2/t) &= HA\hat{x}(t+1/t) + HBu(t-h+1), \\
&\;\;\vdots \\
\hat{y}(t+h+1/t) &= HA^h\hat{x}(t+1/t) + HA^{h-1}Bu(t-h+1) \\
&\qquad\qquad + ... + HBu(t), \\
&\;\;\vdots \\
\hat{y}(t+h+p+1/t) &= \qquad HA^{h+p}\hat{x}(t+1/t) \\
&\quad + HA^{h+p-1}Bu(t-h+1) \\
&\quad + ... + HBu(t+p).
\end{aligned}\right\} \tag{4.6}$$

Comparing (4.6) with (4.1),(4.2), it may be noted that the disturbance noise vector is not included in (4.6) since it is assumed to be unknown. While it may be straightforward to include the disturbance term in the prediction formulation if it is measurable, it is omitted here for the purpose of simplifying the presentation. A composite form of the equation (4.6) may be written as

$$\hat{Y} = G\hat{x}(t+1/t) + F_1\bar{U}_t + F_2 U_t, \tag{4.7}$$

where

$$\hat{Y} = [\hat{y}(t+1/t) \ \hat{y}(t+2/t)...\hat{y}(t+h+p+1/t)]^T,$$

$$G = \begin{bmatrix} H \\ HA \\ \vdots \\ HA^h \\ \vdots \\ HA^{h+p} \end{bmatrix}, \quad F_1 = \begin{bmatrix} 0 & 0 & ... & 0 \\ HB & 0 & ... & 0 \\ \vdots & \vdots & ... & \vdots \\ HA^{h-1}B & HB & ... & 0 \\ \vdots & \vdots & ... & \vdots \\ HA^{h+p-1}B & HA^{h+p-2}B & ... & HA^{p+1}B \end{bmatrix},$$

$$\bar{U}_t = [u(t-h+1), u(t-h+2), ..., u(t-1)]^T,$$

$$F_2 = \begin{bmatrix} 0 & 0 & ... & 0 \\ 0 & 0 & ... & 0 \\ & \vdots & & \\ HB & 0 & ... & 0 \\ & \vdots & & \\ HA^pB & HA^{p-1}B & ... & HB \end{bmatrix},$$

$$U_t = [u(t), u(t+1), ..., u(t+p)]^T.$$

The main difficulty associated with a GPC design based on (4.7) is due to the number of unknowns p corresponding to the number of values in the control sequence $u(t+j)$. One way of reducing the number of unknowns is to predetermine the form of the control sequence. To this end, the control sequence can be made to be constant over the prediction interval,

$$u(t) = u(t+1) = u(t+2) = ... = u(t+p). \tag{4.8}$$

The prediction equation at $t+h+p+1$ horizon can thus be found, as

$$\hat{Y} = G\hat{x}(t+1) + F_1\bar{U}_t + \bar{F}_2\Delta u(t), \tag{4.9}$$

where

$$\bar{F}_2 = \begin{bmatrix} 0 \\ 0 \\ \vdots \\ HB \\ \vdots \\ HA^pB + HA^{p-1}B + ... + HB \end{bmatrix}. \tag{4.10}$$

Since $\hat{x}$ is unknown, it is necessary to estimate it using an observer. In the next section, a Kalman filter is used to address this problem.

4.1.3 Kalman Filter as an Observer

In order to obtain $\hat{x}(t+1)$ from the measured data, a Kalman filter of the following form is considered:

$$\hat{x}(t+1) = A\hat{x}(t) + Bu(t-h) + K(t)(y(t) - H\hat{x}(t)), \qquad (4.11)$$
$$\hat{x}(t_0) = E\{x(t_0)\}. \qquad (4.12)$$

Let

$$e(t) = x(t) - \hat{x}(t), \qquad (4.13)$$

and

$$P(t) = E\{e(t)e^T(t)\}. \qquad (4.14)$$

In the subsequent development, the notation $\hat{x}(t)$ for $\hat{x}(t/t-1)$ will be used. Subtracting (4.11) from (4.1), it can be seen that the reconstruction error $e(t)$ satisfies the following rule:

$$e(t+1) = (A - K(t)H)e(t) + (D - K(t))\xi(t). \qquad (4.15)$$

Following the approach of Goodwin and Sin (1984), the optimal gain matrix $K(t)$ is obtained as

$$K(t) = (AP(t)H^T + D\zeta)(HP(t)H^T + \zeta)^{-1}. \qquad (4.16)$$

The corresponding value of $P(t)$ satisfies the following recursive equation:

$$P(t+1) = (A - K(t)H)P(t)A^T + DD^T\zeta - K(t)D^T\zeta, \qquad (4.17)$$

with $P(t_0) = P_0$ for some $t_0 < t$. Since GPC uses constant feedback, the steady-state value of $K^*(t)$ and $P^*(t)$ may be used as given in the following theorem.

Theorem 4.1 (Goodwin and Sin, 1984)

If (A, D) is stabilisable, (H, A) is detectable, and $P_0 \geq 0$, then $\lim_{t\to\infty} P(t) = P$ where $P(t)$ is the solution of the Riccati difference equation (4.17) with initial condition P_0, P is the unique solution of the algebraic Riccati equation, obtained from the solution of

$$P - APA^T + APH^T(HPH^T + \zeta)^{-1}HPA^T - DD^T\zeta = 0, \qquad (4.18)$$

and $A - KH$ is asymptotically stable, where

$$K = APH^T(HPH^T + \zeta)^{-1}. \qquad (4.19)$$

Throughout the rest of the chapter, the gain matrix $K(t)$ of the filter will be simply denoted as K.

Combining (4.11) with (4.9) yields

$$\hat{Y} = G(A - KH)\hat{x}(t) + GBu(t - h) + F_1\bar{U}_t + \bar{F}_2\Delta u(t)$$
$$+GKy(t). \qquad (4.20)$$

The equation (4.20) may also be rewritten as

$$\hat{Y} = G(A - KH)\hat{x}(t) + GKy(t) + \bar{F}_1\bar{\bar{U}}_t + \bar{F}_2\Delta u(t), \qquad (4.21)$$

where

$$\bar{F}_1 = [GB \quad F_1 + \bar{F}_2 \times [0 \; 0...1]], \qquad (4.22)$$
$$\bar{\bar{U}}_t = [u(t - h) \; u(t - h + 1) \; ...u(t - 1)]^T. \qquad (4.23)$$

4.1.4 Performance Criterion

The following selection of weighting vectors in the performance index (4.5) is considered:

$$R_j = 0, \quad R_0 = r, \qquad (4.24)$$

for $j = 1, ..., p + h$. Given this selection, the performance index reduces to

$$J = \frac{1}{2} \sum_{j=1}^{h+p+1} [y_d(t + j) - \hat{y}(t + j)]^T Q_j[y_d(t + j) - \hat{y}(t + j)]$$
$$+\frac{1}{2}r\Delta u(t)^2 R_j\Delta u(t + j). \qquad (4.25)$$

Substituting (4.21) into (4.25), the solution minimising the performance index may then be obtained by solving

$$\frac{\partial J}{\partial \Delta u(t)} = 0, \tag{4.26}$$

from which direct computations may be obtained explicitly as

$$\Delta u(t) = (\bar{F}_2^T \bar{F}_2 + r)^{-1} \bar{F}_2^T [Y_d - G(A - KH)\hat{x}(t) - \bar{F}_1 \bar{\bar{U}}_t \\ - GKy(t)], \tag{4.27}$$

where

$$Y_d = \begin{bmatrix} y_d(t+1) \\ y_d(t+2) \\ \vdots \\ y_d(t+p+1) \end{bmatrix}. \tag{4.28}$$

Figure 4.1 provides a graphical illustration of this scheme.

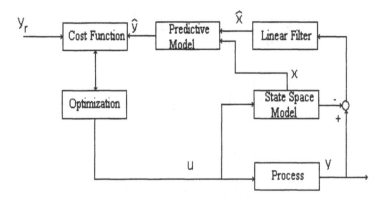

Fig. 4.1. GPOC control scheme

4.2 Stability and Robustness Analysis

In this section, the stability and robustness of the predictive control law derived in the preceding section will be analysed. The analysis involves two parts: (1) by using a perturbed theory, the stability of the closed-loop system is studied; and (2) for sufficiently small perturbation of the model parameters, the robustness is guaranteed using a detailed mathematical analysis.

4.2.1 Stability

For the GPOC designed as described, it is interesting to examine the closed-loop stability properties of the system. In the following development, a sufficient condition pertaining to stability is provided which shows that stability may be ensured with an appropriate choice of prediction horizon and weighting matrices.

Substituting (4.13) into (4.27) produces

$$\Delta u(t) = (\bar{F}_2^T \bar{F}_2 + r)^{-1} \bar{F}_2^T [Y_d - GAx(t) + G(A - KH)e(t)$$
$$- \bar{F}_1 \bar{\bar{U}}_t - GK\xi(t)]. \tag{4.29}$$

(4.29) may be rewritten as:

$$\begin{bmatrix} u(t-h+1) \\ u(t-h+2) \\ \vdots \\ u(t-1) \\ u(t) \end{bmatrix} = \begin{bmatrix} 0\,1\,...\,0 \\ 0\,0\,...\,0 \\ \vdots\,\vdots\,\vdots\,\vdots \\ 0\,0\,...\,1 \\ 0\,0\,...\,\epsilon r \end{bmatrix} \begin{bmatrix} u(t-h) \\ u(t-h+1) \\ \vdots \\ u(t-2) \\ u(t-1) \end{bmatrix}$$

$$-\epsilon \begin{bmatrix} O_{(h-1)\times h} \\ \bar{F}_2^T GB\,\bar{F}_2^T F_1 \end{bmatrix} \begin{bmatrix} u(t-h) \\ u(t-h+1) \\ \vdots \\ u(t-1) \end{bmatrix}$$

$$-\epsilon \begin{bmatrix} O_{(h-1)\times n} \\ \bar{F}_2^T GA \end{bmatrix} x(t)$$

$$+\epsilon \begin{bmatrix} O_{(h-1)\times n} \\ \bar{F}_2^T G(A-KH) \end{bmatrix} e(t) + \epsilon \begin{bmatrix} O_{(h-1)\times(h+p)} \\ \bar{F}_2^T \end{bmatrix} Y_d$$

$$-\epsilon \begin{bmatrix} O_{(h-1)\times 1} \\ \bar{F}_2^T GK \end{bmatrix} \xi(t), \tag{4.30}$$

where

$$\epsilon = (\bar{F}_2^T \bar{F}_2 + r)^{-1}.$$

Let

$$\Theta(t) = [u(t-h)\,u(t-h+1)...u(t)]^T,$$

$$A_u = \begin{bmatrix} 0 & 1 & \dots & 0 \\ 0 & 0 & \dots & 0 \\ \vdots & \vdots & \vdots & \vdots \\ 0 & 0 & \dots & 1 \\ 0 & 0 & \dots & 0 \end{bmatrix},$$

$$A_{u1} = \begin{bmatrix} 0 & 0 & \dots & 0 \\ 0 & 0 & \dots & 0 \\ \vdots & \vdots & \vdots & \vdots \\ 0 & 0 & \dots & 0 \\ 0 & 0 & \dots & 1 \end{bmatrix},$$

$$A_{u2} = \begin{bmatrix} O_{(h-1)\times h} \\ \bar{F}_2^T GB \ \bar{F}_2^T F_1 \end{bmatrix},$$

$$A_x = \begin{bmatrix} O_{(h-1)\times n} \\ \bar{F}_2^T GA \end{bmatrix},$$

$$A_e = \begin{bmatrix} O_{(h-1)\times n} \\ \bar{F}_2^T G(A - KH) \end{bmatrix},$$

$$A_y = \begin{bmatrix} O_{(h-1)\times(h+p)} \\ \bar{F}_2^T \end{bmatrix},$$

$$A_\xi = \begin{bmatrix} O_{(h-1)\times 1} \\ \bar{F}_2^T GK \end{bmatrix}.$$

Thus,

$$\Theta(t+1) = (A_u + \epsilon r A_{u1} - \epsilon A_{u2})\Theta(t) - \epsilon A_x x(t) + \epsilon A_e e(t)$$
$$+ \epsilon A_y Y_d - \epsilon A_\xi \xi(t).$$

Combining (4.30),(4.1) with (4.15) yields

$$\begin{bmatrix} x(t+1) \\ \Theta(t+1) \\ e(t+1) \end{bmatrix} = \begin{bmatrix} A & \bar{B} & 0 \\ -\epsilon A_x & A_u + \epsilon A_{u1} - \epsilon A_{u2} & \epsilon A_e \\ 0 & 0 & A - KH \end{bmatrix} \begin{bmatrix} x(t) \\ \Theta(t) \\ e(t) \end{bmatrix}$$
$$+ \epsilon \begin{bmatrix} 0 \\ A_y \\ 0 \end{bmatrix} Y_d + \begin{bmatrix} D \\ -\epsilon \bar{F}_2^T GK \\ D - K \end{bmatrix} \xi(t), \qquad (4.31)$$

where $\bar{B} = B[1 \ 0 \ 0...0]$. Theorem 4.2 follows directly.

Theorem 4.2

The system (4.1) with the control law (4.27) is stable if it can be guaranteed that all the eigenvalues of

$$\begin{bmatrix} A & \bar{B} & 0 \\ -\epsilon A_x & A_u + \epsilon A_{u1} - \epsilon A_{u2} & \epsilon A_e \\ 0 & 0 & A - KH \end{bmatrix} \tag{4.32}$$

are within the unit circle by appropriately choosing the prediction horizon p and the weighting value r.

Let

$$Z(k) = \begin{bmatrix} x(t) \\ \Theta(t) \\ e(t) \end{bmatrix}, \quad \Lambda = \begin{bmatrix} A & \bar{B} & 0 \\ 0 & A_u & 0 \\ 0 & 0 & A - KH \end{bmatrix},$$

$$E_1 = \begin{bmatrix} 0 & 0 & 0 \\ A_x & A_{u1} & A_e \\ 0 & 0 & 0 \end{bmatrix}, \quad E_2 = \begin{bmatrix} 0 & 0 & 0 \\ 0 & A_{u1} & 0 \\ 0 & 0 & 0 \end{bmatrix}.$$

Then, the free system of (4.31) can be rewritten as

$$Z(t+1) = \Lambda Z(t) + \epsilon (E_1 + r E_2) Z(t). \tag{4.33}$$

Define

$$P_{\Lambda E_1} = ||\Lambda^T P E_1 + E_1^T P \Lambda||,$$
$$P_{\Lambda E_2} = ||\Lambda^T P E_2 + E_2^T P \Lambda||,$$
$$P_{E_1 E_2} = ||E_1^T P E_2 + E_2^T P E_1||,$$
$$P_{E_1^2} = ||E_1^T P E_1||,$$
$$P_{E_2^2} = ||E_2^T P E_2||,$$

where P is the solution to the following equation:

$$\Lambda^T P \Lambda - P = -2I, \tag{4.34}$$

and I is the unit matrix.

Lemma 4.1 (Horn and Johnson, 1985)

Let A, B be Hermitian matrices and let the eigenvalues $\lambda_i(A), \lambda_i(B)$ and $\lambda_i(A + B)$ be arranged in increasing order, i.e., $\lambda_{min} = \lambda_1 \leq \lambda_2 \leq ... \leq \lambda_n = \lambda_{max}$. For each $k = 1, 2, ..., n$,

$$\lambda_k(A) + \lambda_1(B) \leq \lambda_k(A + B) \leq \lambda_k(A) + \lambda_n(B). \tag{4.35}$$

Using Lemma 4.1, if r or p are chosen properly, the following theorem ensures the stability of the system.

Theorem 4.3

Let the system be described by (4.1) and controlled by the control law (4.27). Assume the system is open-loop stable. Then, the free system of (4.31) can be stabilised if r and p satisfy

$$\delta = \max\{\epsilon, \epsilon r, \epsilon r^2\}$$

$$< \min\left\{1, \frac{2}{P_{\Lambda E_1} + P_{\Lambda E_2} + P_{E_1 E_2} + P_{E_1^2} + P_{E_2^2}}\right\}. \tag{4.36}$$

Proof

While the open-loop system of the system (4.1) is asymptotically stable, it follows that Λ in (4.31) will be asymptotically stable, since A_u and $A - KH$ are also asymptotically stable. Since Λ is asymptotically stable, the solution of P in (4.34) is positive definite. Let

$$V(Z(t)) = Z^T(t)PZ(t). \tag{4.37}$$

The derivative of $V(Z(t))$ satisfies

$$\begin{aligned}
\Delta V(Z(t)) &= Z^T(t+1)PZ(t+1) - Z(t)^T PZ(t), \\
&= Z^T(t)\{[\Lambda + \epsilon(E_1 + rE_2)]^T P[\Lambda + \epsilon(E_1 + rE_2)] - P\}Z(t), \\
&= Z^T(t)\{\Lambda^T P\Lambda - P + \epsilon[\Lambda^T P(E_1 + rE_2) + (E_1 + rE_2)^T P\Lambda] \\
&\quad + \epsilon^2(E_1 + rE_2)^T P(E_1 + rE_2)\}Z(t), \\
&= Z^T(t)\{-2I + \epsilon[\Lambda^T P(E_1 + rE_2) + (E_1 + rE_2)^T P\Lambda] \\
&\quad + \epsilon^2(E_1 + rE_2)^T P(E_1 + rE_2)\}Z(t). \tag{4.38}
\end{aligned}$$

By using Lemma 4.1,

$$\begin{aligned}
\Delta V(Z(t)) &\leq Z^T(t)\{-2 + \lambda_{max}[\epsilon(\Lambda^T PE_1 + E_1^T P\Lambda) \\
&\quad + \epsilon r(\Lambda^T PE_2 + E_2^T P\Lambda) \\
&\quad + \epsilon^2(E_1 + rE_2)^T P(E_1 + rE_2)]\}Z(t).
\end{aligned}$$

$$\tag{4.39}$$

Thus, if

$$\lambda_{max}[\epsilon(\Lambda^T P E_1 + E_1^T P\Lambda) + \epsilon r(\Lambda^T P E_2 + E_2^T P\Lambda)$$
$$+\epsilon^2(E_1 + rE_2)^T P(E_1 + rE_2)] < 2, \qquad (4.40)$$

then $\Delta V(Z(t)) < 0$. It is noted that

$$\lambda_{max}[\epsilon(\Lambda^T P E_1 + E_1^T P\Lambda) + \epsilon r(\Lambda^T P E_2 + E_2^T P\Lambda)$$
$$+ \epsilon^2(E_1 + rE_2)^T(E_1 + rE_2)] \le \epsilon[||\Lambda^T P E_1 + E_1^T P\Lambda||$$
$$+ r||\Lambda^T P E_2 + E_2^T P\Lambda|| + \epsilon(||E_1^T P E_1||$$
$$+ r^2||E_2^T P E_2|| + r||E_1^T P E_2 + E_2^T P E_1||)].$$

Assume $\epsilon < 1$. Then, it is easy to show that

$$\epsilon[P_{\Lambda E_1} + rP_{\Lambda E_2} + \epsilon(P_{E_1^2} + r^2 P_{E_2^2} + rP_{E_1 E_2})]$$
$$< \epsilon(P_{\Lambda E_1} + rP_{\Lambda E_2} + P_{E_1^2} + r^2 P_{E_2^2} + rP_{E_1 E_2})$$
$$\le max\{\epsilon, \epsilon r, \epsilon r^2\}(P_{\Lambda E_1} + P_{\Lambda E_2}$$
$$+ P_{E_1^2} + P_{E_2^2} + P_{E_1 E_2}).$$

Therefore, the condition (4.40) can be satisfied if (4.36) holds. The theorem is proven.

Remark 4.1

The main problem addressed in Theorem 4.3 is to determine r, p in (4.36) in order to maintain the stability of the closed-loop system. There are some degrees of freedom for this problem: the choice of the r and p. It would be interesting to take advantage of these features in order to optimise the robustness of the closed-loop system. For given Q, R, the p, r that yield maximal robustness are solutions of the following optimisation problem:

$$\max_{p,r} \delta = \{\epsilon, \epsilon r, \epsilon r^2\}$$

$$\text{subject to} \quad \delta < min\left\{1, \frac{2}{P_{\Lambda E_1} + P_{\Lambda E_2} + P_{E_1 E_2} + P_{E_1^2} + P_{E_2^2}}\right\}.$$

This problem is indeed a difficult one. In the general case, there is no analytical solution. Non-linear optimisation techniques can be used to cope with this problem and to obtain a sub-optimal solution.

Theorem 4.4

Let the system be described by (4.1),(4.2) and controlled by the control law (4.27). Assume the system is open-loop stable, Theorem 4.3 holds, and $y_d(t+1), y_d(t+2)...y_d(t+h+p+1)$ are bounded. Then, the system (4.1),(4.2) with the control law (4.27) is stable, *i.e.*, $\{y(t)\}$ and $\{u(t)\}$ are bounded for all t.

Proof

Since Y_d and $\xi(t)$ are bounded and the free system is asymptotically stable from Theorem 4.3, the conclusion follows from the result of Payne (1987).

4.2.2 Robustness

Thus far, the nominal stability of the system has been considered. Under general modelling perturbations, it is useful to examine the robustness properties of the closed-loop system. Next, consider the stability of the controlled system (4.1) and (4.27) with the real parameters perturbed to A_r, B_r, H_r and h_r.

Define

$$A_r = A + E_A, B_r = B + E_B, H_r = H + E_H. \tag{4.41}$$

However, in a practical situation, one would not know exactly the perturbation matrices. One may only have knowledge of the magnitude of the maximum deviation that can be expected in the entries of each perturbation matrix. In this case, the entries of each perturbation matrix may be such that

$$|E_{A_{ij}}| < \eta, |E_{B_{ij}}| < \eta, |E_{H_{ij}}| < \eta, \tag{4.42}$$

where η is the magnitude of the maximum deviation. The robustness of the system (4.1),(4.27) under small parameter perturbations will be studied.

Since the pure time delay h is an integer, this implies h can be measured as an integer multiple of the sampling period. Thus, when $|h_r - h|$ is small enough, $h_r = h$. Therefore, the system perturbation will mainly involve A_r, B_r and H_r.

The perturbed system is described by

$$x(t+1) = (A + E_A)x(t) + (B + E_B)u(t-h) + D\xi(t), \tag{4.43}$$
$$y(t) = (H + E_H)x(t) + \xi(t). \tag{4.44}$$

The error equation is obtained from (4.43), (4.44) and (4.11).

$$e(t+1) = (A - KH)e(t) + (E_A - KE_H)x(t) + E_Bu(t-h)$$
$$+(D - KH)\xi(t). \qquad (4.45)$$

As a similar procedure to (4.31), the system (4.1),(4.2),(4.27) with parameter perturbation (4.41) is given in the following form:

$$X(t) = \bar{A}X(t) + \Delta\bar{A}X(t) + \begin{bmatrix} 0 \\ \epsilon\bar{F}_2^T G \\ 0 \end{bmatrix} Y_d + \begin{bmatrix} D \\ -\epsilon\bar{F}_2^T GD \\ D - K \end{bmatrix} \xi(t), \qquad (4.46)$$

where

$$X(t) = [x^T(t+1) \ \Theta^T(t+1) \ e^T(t+1)]^T, \qquad (4.47)$$

$$\bar{A} = \begin{bmatrix} A & B & 0 \\ -\epsilon A_x & A_u + \epsilon r A_{u1} + \epsilon A_{u2} & \epsilon A_e \\ 0 & 0 & A - KH \end{bmatrix}, \qquad (4.48)$$

$$\Delta\bar{A} = \begin{bmatrix} E_A & \bar{E}_B & 0 \\ 0 & 0 & 0 \\ E_A - KE_H & \bar{E}_B & 0 \end{bmatrix}, \qquad (4.49)$$

and $\bar{E}_B = E_B[1 \ 0 \ 0...0]$.

Throughout the rest of this chapter, $||[\cdot]||$ will be used to represent the modulus matrix, i.e., a matrix with modulus entries. $([\cdot])_s$ represents the symmetric part of the matrix $[\cdot]$ and $\rho(\cdot)$ denotes the spectral radius. If $A \in R^{n \times n}$, $\rho(A) = max\{|\lambda| : \lambda \in \lambda(A)\}$.

Define

$A \geq B$ if all $a_{ij} \geq b_{ij}$, where a_{ij}, b_{ij} are the elements of the matrices A, B, respectively.

Lemma 4.2 (Horn and Johnson, 1985)

If $|A| \leq B$, then $\rho(A) \leq \rho(|A|) \leq \rho(B)$.

Theorem 4.5

Assume that the nominal system (4.1) is stabilised by Theorem 4.2 or Theorem 4.3. If the system parameter perturbations are sufficiently small, the control system in (4.43) with the control law (4.27) is robustly stable.

Proof

Since the asymptotic stability of the nominal system (noting that p and r are fixed) can be guaranteed, it is known that $\bar{A}$ is asymptotically stable. It is well known that for a given matrix $Q > 0$, such as $Q = 2I$, the solution P of the following Lyapunov equation

$$\bar{A}^T P \bar{A} - P = -2I,$$

where I is the unit matrix, is positive definite. Hence,

$$V(X(k)) = X^T(k)PX(k),$$

is a positive definite function. Then, it can be shown that

$$
\begin{aligned}
\Delta V(X(t)) &= X^T(k+1)PX(k+2) - X(k)^T PX(k), \\
&= X^T(k)[\bar{A}^T P\bar{A} - P + \bar{A}^T P\Delta\bar{A} + \bar{A}^T P\bar{A} \\
&\quad + \Delta\bar{A}^T P\Delta\bar{A}]X(k), \\
&= X^T(k)[-2I + \bar{A}^T P\Delta\bar{A} + \bar{A}^T P\bar{A} \\
&\quad + \Delta\bar{A}^T P\Delta\bar{A}]X(k).
\end{aligned}
\tag{4.50}
$$

Let

$$M = -2I + \bar{A}^T P\Delta\bar{A} + \bar{A}^T P\bar{A} + \Delta\bar{A}^T P\Delta\bar{A}. \tag{4.51}$$

For the perturbed system in (4.43),(4.27) to be asymptotically stable, $\Delta V(x)$ is negative definite, or equivalently, all the eigenvalues of M are negative, $i.e.$,

$$\lambda(M) < 0. \tag{4.52}$$

In view of Lemma 4.2,

$$\lambda_{max}(M) \leq -2 + \lambda_{max}[\bar{A}^T P\Delta\bar{A} + \bar{A}^T P\bar{A} + \Delta\bar{A}^T P\Delta\bar{A}]. \tag{4.53}$$

Thus, if

$$\lambda_{max}[\bar{A}^T P\Delta\bar{A} + \bar{A}^T P\bar{A} + \Delta\bar{A}^T P\Delta\bar{A}] < 2, \tag{4.54}$$

then (4.52) holds. Note that

$$\lambda_{max}[\bar{A}^T P \Delta \bar{A} + \bar{A}^T P \bar{A} + \Delta \bar{A}^T P \Delta \bar{A}] \le \rho[\bar{A}^T P \Delta \bar{A} + \bar{A}^T P \bar{A}$$
$$+\Delta \bar{A}^T P \Delta \bar{A}].$$

Since

$$\bar{A}^T P \Delta \bar{A} + \bar{A}^T P \bar{A} + \Delta \bar{A}^T P \Delta \bar{A} \le 2\eta(|\bar{A}||P|U_{\bar{A}})_s + \eta^2 U_{\bar{A}}^T |P|U_{\bar{A}}, \quad (4.55)$$

where $U_{\bar{A}}$ is an $n \times n$ matrix with $U_{\bar{A}_{ij}} = 1$ for all $i, j = 1, 2, ..., n$, applying Lemma 4.2

$$\rho[\bar{A}^T P \Delta \bar{A} + \bar{A}^T P \bar{A} + \Delta \bar{A}^T P \Delta \bar{A}] \le \rho[2\eta(|\bar{A}||P|U_{\bar{A}})_s$$
$$+\eta^2 U_{\bar{A}}^T |P|U_{\bar{A}}], \quad (4.56)$$

Since η is small enough, one has $\eta < 1$ which yields

$$\rho[2\eta(|\bar{A}||P|U_{\bar{A}})_s + \eta^2 U_{\bar{A}}^T |P|U_{\bar{A}}] \le \eta \rho[2(|\bar{A}||P|U_{\bar{A}})_s + U_{\bar{A}}^T |P|U_{\bar{A}}]. \quad (4.57)$$

It can readily be shown that, if

$$\eta \rho[2(|\bar{A}||P|U_{\bar{A}})_s + U_{\bar{A}}^T |P|U_{\bar{A}}] < 2, \quad (4.58)$$

then (4.52) is satisfied. Therefore, if η is sufficiently small, the inequality (4.58) always holds and then the free system of (4.46) is stable. Since Y_d and $\xi(t)$ are bounded, following Payne (1987), the theorem is proven.

4.3 Simulation Study

Injection moulding is one of the major processes used in the plastics manufacturing industry. Control of the injection moulding process involves three phases: filling, packing and holding, and cooling. This application example is mainly concerned with the process control of the filling phase.

The actuator used to control injection speed is typically a proportional control hydraulic valve. The input is an electrical voltage applied to the proportional control valve. By varying the input voltage at different time intervals through an effective algorithm, the ram velocity can be controlled to a predetermined velocity profile.

A fourth-order model reported by Wang (1984) is used, relating the servovalve opening to the injection ram velocity, and described by

$$G_P(s) = \frac{2.144 \times 10^{14}}{(s+125)(s+1138)[(s+383)^2 + 1135^2]}. \qquad (4.59)$$

To apply the GPC, a discrete-time model related to the servovalve opening to the injection ram velocity in the z-domain may be obtained directly from (4.59) with zero-order hold and sampling time $T = 0.001$s:

$$
\begin{aligned}
G(z) &= \frac{C(z)}{M(z)} \\
&= \frac{0.0058z^{-1} + 0.040z^{-2} + 0.027z^{-3} + 0.0017z^{-4}}{1 - 1.78z^{-1} + 1.44z^{-3} - 0.72z^{-3} + 0.13z^{-4}} z^{-h}. \qquad (4.60)
\end{aligned}
$$

Furthermore, Zhang (1996) and Agrawal (1987) suggested associating the model with an input delay of 25ms to model the time delay in the transmission of oil flow signals.

Thus, it is written in a state space form from (4.60) with the presence of input delay

$$
\begin{aligned}
x(t+1) &= \begin{bmatrix} 1.78 & 1 & 0 & 0 \\ -1.44 & 0 & 1 & 0 \\ 0.72 & 0 & 0 & 1 \\ -0.13 & 0 & 0 & 0 \end{bmatrix} x(t) + \begin{bmatrix} 0.0058 \\ 0.040 \\ 0.027 \\ 0.0017 \end{bmatrix} u(t-h) \\
&\quad + \begin{bmatrix} 1.88 \\ -1.34 \\ 0.82 \\ -0.03 \end{bmatrix} \xi, \\
y &= [1\ 0\ 0\ 0] + \xi(t),
\end{aligned}
$$

where h is 25 sampling intervals, $y(t)$ and $u(t)$ are the ram velocity of the injection moulding machine and the control valve opening at a discrete time t, respectively, and $\xi(t)$ is an uncorrelated random sequence with zero mean and variance ζ representing the process noise.

The entire phase may be regulated by controlling the injection speed of the ram so as to follow a pre-generated trajectory. Figure 4.2 shows a typical ram velocity profile to fill a mould of increasing cross-section initially and having a constant cross-section thereafter. The profile for this part can be divided into several phases. During Phase I of the figure, which corresponds to filling the runner, velocity is very rapidly increased and then held constant. During this time, the concern is for minimum injection time and heat loss in the material. In Phase II, the velocity is rapidly reduced to eliminate jetting at the gate. Once the melt enters the cavity, the aim is to maintain a constant flow front speed. Phase III corresponds to filling the increasing cross-section

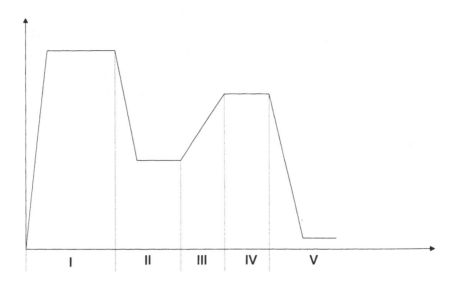

Fig. 4.2. Ram velocity profile

portion of the cavity while Phase IV is concerned with filling the constant cross-section portion of the cavity. The ram velocity profile should increase to maintain constant velocity of the melt across the mould surface during Phase III, while the ram velocity should be maintained at a constant value during Phase IV. Finally. the velocity is reduced in Phase V to eliminate flashing and/or overpacking. All of the above objectives have to be achieved in the short time available for mould filling.

The simulation chart is shown in Fig. 4.3. T_{termin} represents the simulation time of a control cycle. In order to demonstrate the effectiveness of the control scheme a proportional-integral (PI) controller is also applied to the same model with the same sampling period. First, to illustrate the performance of the controller, a comparison is made with the proportional-integral-derivative (PID) type controller (Smith and Corripio, 1985). This is a typical PI controller for a linear system with delay according to the optimal integral absolute error (IAE) tuning criterion. For a fair comparison, the same process with the same control performance will be used in the simulation study, where the control performance is measured by the summation of set-point tracking errors and control actions. The weightings are chosen as $p = 10, Q = 1$, and $r = 170$ to achieve the same performance as the PID-type controller. Figure 4.4 shows that the control scheme has a superior control performance over the PI controller of Smith and Corripio (1985), when both have almost the same performance index 800.3 over the simulation time interval. Notice that no noise has been introduced in this simulation.

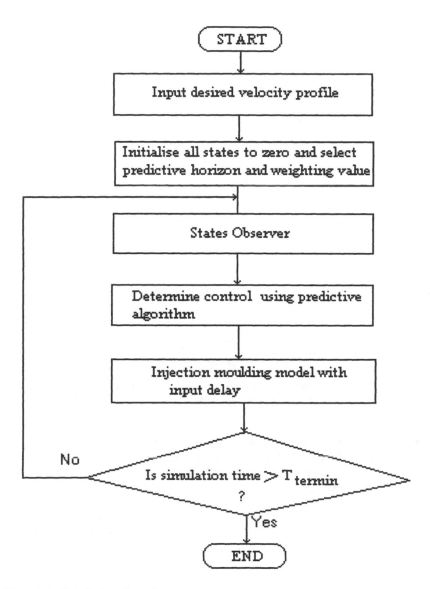

Fig. 4.3. Simulation flow chart

Further, to see the performance in the face of process and measurement noise, the GPOC parameters are selected as $p = 10, Q = 1$, and $r = 85$. It can be checked that $|\lambda| < 1$ using Theorem 4.2. Thus, the controller can guarantee the stability of the closed-loop system. In the following simulation, all the initial values of the system states chosen are zero while all the initial values of the system observers are taken to be 0.01.

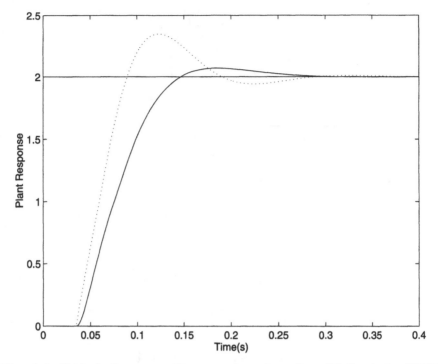

Fig. 4.4. Control of process with zero noise, where the solid line is for GPOC control while the dotted line is for the Smith and Corripio (1985) PI controller

Figure 4.5 shows the system responses and the corresponding controller output over 1000 sampling intervals. Measurement noise of $\zeta = 0.05$ is introduced in the simulation to test the noise rejection capabilities. The state and observer trajectories are plotted in Fig. 4.5.

Next, ζ is set to 0.1 and the controller performance is tested in a similar way. The results are shown in Fig. 4.6. It can be seen from these figures that the controller in conjunction with the observers exhibit good control performance in the presence of both time delay and noise.

To compare the performance of the method with that of a PID-type controller, a PI controller with tuning algorithm in accordance with Smith and Corripio (1985) is employed for ram velocity control. The PI controller is

$$u(t) = k_p(y_d(t) - y(t)) + k_i \sum_{i=0}^{k-1}(y_d(i) - y_i(t)), \tag{4.61}$$

with $k_p = 0.0775$ and $k_i = 0.0082$. The system responses for noise $\zeta = 0.05$ and $\zeta = 0.1$ for the PI controller are plotted in Fig. 4.7. Comparing Fig.

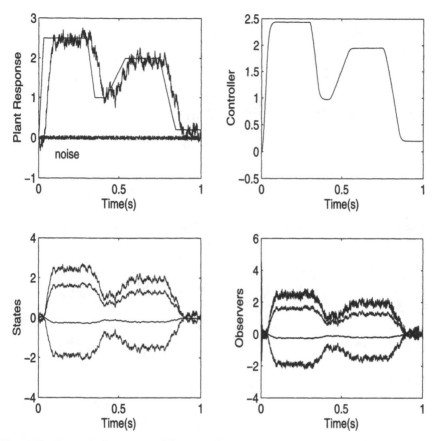

Fig. 4.5. Control of process with noise $\zeta = 0.05$

4.7 with Fig. 4.5 and Fig. 4.6, the system response of the PI controller has a much larger deviation from the reference profiles than that of the control, and the control input of the new controller has a smoother action than the Smith and Corripo (1985) PI controller.

The control algorithm is robustly stable against small perturbation of the system parameters. For example, when a_1 changes from 1.7805 to 1.78, the control result is shown in Fig. 4.8. When perturbation of the model parameters is large, it is recommended to use the recursive least squares method to identify the parameters on-line, and then for the predictive controller to adopt the updated parameters at every sampling instant. For instance, it is assumed that the model parameters vary as shown in Fig. 4.9. Figure 4.10 shows the control result of the predictive control together with the recursive least squares method. Clearly, the control can achieve satisfactory tracking performance against perturbation of the model parameters.

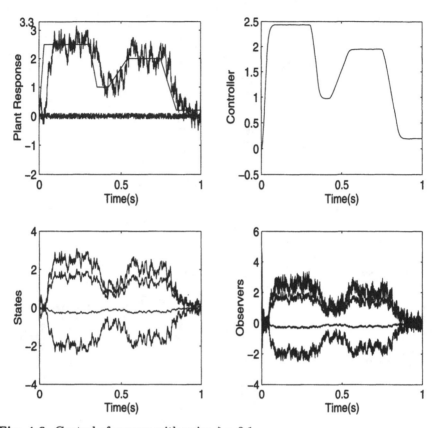

Fig. 4.6. Control of process with noise $\zeta = 0.1$

For practical application control, there are two tuning when applying the controller: one is the prediction horizon p, and one is the control weighting r. In general, most researchers suggest that the prediction horizon p may be set between 10 and 20. Here, it is recommended that $p = 10$. Now only parameter r affects the control response. From Theorem 4.1, it can be seen that a large value of r ensures system stability but reaching steady state takes a long time. For instance, when selecting control weightings of $r = 0.01$, and 15, the control responses are as shown in Fig. 4.11. Obviously, a small value of r improves the speed of response but the stability is not as good as that for control weighting $r = 15$. To make the trade-off between response time and stability, a tuning facility on the control board can be provided. This adjustment can be set manually to follow the desired set-point according to the response curve.

Remark 4.2

Since the controller is based on the model of the system, it is important to

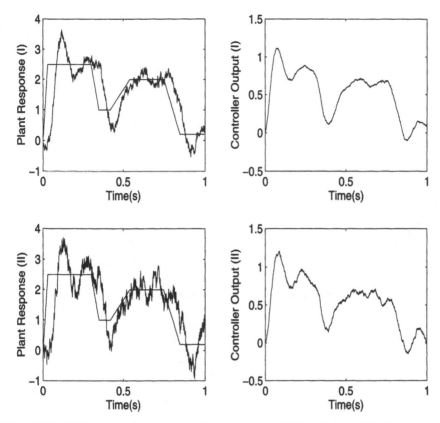

Fig. 4.7. PI Control of process with noise $\zeta = 0.05$ and $\zeta = 0.1$: the system response (I) representing the system output for noise $\zeta = 0.05$ while the system response (II) representing the system output for noise $\zeta = 0.1$

know the model parameters. If the model is unavailable, the recursive least squares method should be used to identify the system parameters off-line or on-line.

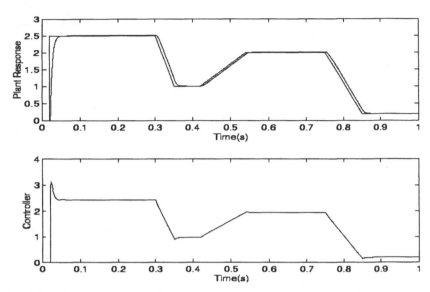

Fig. 4.8. Control response when model parameter a_1 is perturbed

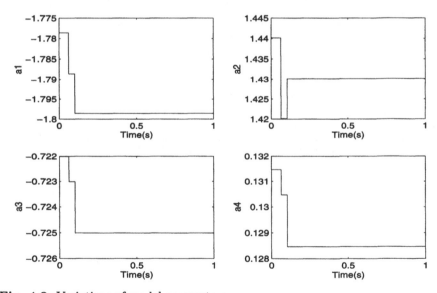

Fig. 4.9. Variations of model parameters

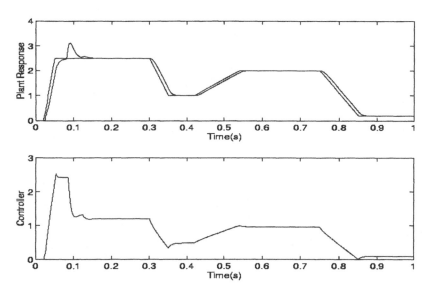

Fig. 4.10. Control response using predictive control with the recursive least squares method

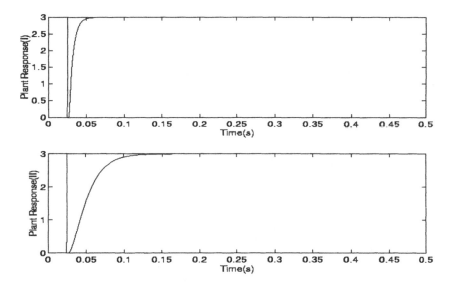

Fig. 4.11. Control responses for different values of r: system response (I) is for $r = 0.01$ while system response (II) is for $r = 15$

CHAPTER 5

OPTIMAL PID CONTROL BASED ON A PREDICTIVE CONTROL APPROACH

Proportional-integral-(derivative) (PI(D)) controllers have remained the most commonly used controllers in industrial process control for more than fifty years, despite the advance in mathematical control theory. The main reason is that they have a simple structure, which is easily to be understood by engineers, and under practical conditions it performs more reliably than more advanced and complex controllers.

Over the years, numerous techniques have been suggested for tuning PID parameters. However, each technique is usually applicable only to a limited class of systems. In Åström and Hägglund (1995), tuning rules are developed for systems with monotonic step responses which may be adequately described using a first-order model with dead time. Most literature reports (Chidambaram, 1994; Hägglund, 1992; Sung et al., 1996) are also based on this model. In Sung et al. (1996), PID tuning for under-damped systems is considered. Systems with integrating action are addressed in Wang and Cluett (1999), and Kaya and Atherton (1999), and those exhibiting non-minimum phase characteristics are addressed in Luyben (1999). Separate attention is focused on unstable systems, with and without dead time in de Paor and O'Malley (1985), Stahl and Hippe (1987), Venkatashankar and Chidambaram (1994), Huang and Chen (1997), Ho and Xu (1998), and Rostein and Lewin (1991). PID control for servomechanisms are dealt with in Tan and Lee (1998) and Huang et al. (2001), with a distinct difference from the same PID applied to system control. Furthermore, different variations of PID control structure are configurable for different systems. For example, for time-delay systems, usually only a PI controller structure is used, since the derivative action invokes an undesirable response. For systems with long time constants or integrating action, a PD structure may be used instead.

With the many classes of PID designs and structures available, it has become necessary to determine the characteristics of the system before the appropriate class of design rules can be selected and applied. Furthermore, apart from the difference in the associated model structure, the basis and nature on which these tuning rules are derived differ from one approach to another. The pre-specifications required for commissioning the control system are accordingly also different.

Another well-known constraint associated with fixed gain conventional PID control is that the control structure rapidly loses effectiveness when applied to systems having complex dynamics, such as those with long dead times, poor damping or unstable dynamics. Under these circumstances, more complex control structures become necessary, such as those using a dead time compensator for time-delay systems. This constraint has remained with PID control since its inauguration.

This chapter presents PI and PID control designs based on the application of generalised predictive control (GPC) to a general linear system model which is representative of many classes of systems encountered in industry (Tan et al., 1998; Tan et al., 1999; Tan et al., 2000). It will be shown that the controller can be implemented with a limited set of instructions, available in most distributed control systems, and that the computation time required, even for tuning, is very short. The method of GPC implementation is based on the fact that a wide range of systems in industry can be described by a simple first-order or second-order plus delay model and that a set of simple Ziegler–Nichols type functions relating controller parameters to system parameters can be obtained. In addition, the design is intended to yield an optimum tracking control performance according to a performance index. The main idea is based on back-calculating an equivalent set of PID parameters from a GPC control law derived using a first-order or second-order general system model. In this way, an optimal controller is developed which has a simple and desirable PID structure, but which can yield a GPC level of performance. One less desirable prerequisite and possible drawback to the application of a GPC solution is the need to select the weight matrices, which may not be easily and effectively accomplished by the average control engineer. To overcome this difficulty, relationships are derived, mapping the weighting parameters to classical pole placement type control specifications in terms of the desired natural frequency and the damping ratio of the closed-loop system. Thus, specifications may be made in terms of more intuitive specifications and an initial set of equivalent GPC weights derived for either further fine tuning, or used directly.

5.1 Structure of the PID Controller

The PID controller uses the following control law:

$$u(t) = K(e(t) + \frac{1}{T_p}\int_0^t e(\tau)d\tau + T_d\frac{de(t)}{dt}). \qquad (5.1)$$

In much of the literature, this is also known as the ideal type of PID control. For the sake of simplicity, the chapter will focus only on this basic structure without considering other minor structural modifications for derivative filtering, set-point weighting *etc.* This type is shown diagrammatically as in Fig. 5.1, and is the one normally described in text books.

The above continuous-time PID can be written in discrete form in the time domain

$$u(t) = Ke(t) + \frac{1}{T_p} \sum_{i=1}^{t} e(i) + T_d[e(t) - e(t-1)]. \qquad (5.2)$$

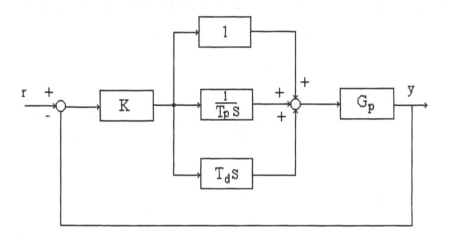

Fig. 5.1. Structure of PID

5.2 Optimal PI Controller Design Based on GPC

In this section, a detailed GPC solution for a first-order system with time delay is presented. The useful properties of GPC control are combined with the simplicity of the PI structure by applying the GPC solution to tune PI control gains for first-order time-delay systems or those which may be adequately modelled as such. It will be proven that the closed-loop system with input delay is stable under the control law. One less desirable prerequisite and possible drawback to the application of the GPC solution is the need to select the weight matrices. This task may be rather abstract, so that selection is not easily and effectively accomplished by the average system operator or control engineer. Based on this consideration, a criterion for selection of the weight matrices is presented with the advantage that the tuning parameters are closely associated with the desired natural frequency and damping ratio of the closed-loop system. The robustness property of the tuned control system is also analysed, and it is shown that the design presented is robustly stable for small modelling perturbations.

5.2.1 GPC Solution for Time-delay Systems

In this section, the GPC solution for a linear system with time delay is provided together with an analysis of the stability of the closed-loop system.

Consider a single-input linear system with time delay described by:

$$x(k+1) = Fx(k) + Bu(k-h), \tag{5.3}$$

where x is an $n \times 1$ state vector, u is a 1×1 vector, F and B are given matrices with proper dimensions, and h is the input of delay time. The problem is to find a series of M control $u(k), ...u(k+M-1)$, at each time $t = kT$, so that the quadratic cost function:

$$J = \sum_{l=1}^{P} \|x(k+l)\|_{Q(l)}^2 + \sum_{j=1}^{M} \|u(k+j-1)\|_{R(j)}^2, \tag{5.4}$$

is minimised, where $Q(l) \geq 0$ is the output weighting matrix, $R(j) \geq 0$ is the control weighting matrix, and P and M ($N \geq P \geq M$) are the prediction and control horizons respectively.

The dynamic system (5.3) may be time-segregated as follows:

$$x(k+1) = Fx(k), \quad 0 \leq k < h, \tag{5.5}$$
$$x(k+1) = Fx(k) + Bu(k-h), \quad k \geq h. \tag{5.6}$$

The future outputs $x(k+l)$ in (5.3) are obtained from a prediction with M control based on the predictive model (5.3)

$$x(k+1) = Fx(k) + Bu(k-h),$$
$$x(k+2) = F^2x(k) + FBu(k-h) + Bu(k-h+1),$$

$$\vdots$$

$$x(k + P - 1) = F^{P-1}x(k) + F^{P-2}Bu(k - h) + \dots$$
$$+F^{P-M-1}Bu(k - h + M - 2),$$
$$x(k + P) = F^{P}x(k) + F^{P-1}Bu(k - h) + \dots$$
$$+F^{P-M}Bu(k - h + M - 1).$$

The equations above can also be rewritten in matrix formulation as

$$Y = GFx(k) + AU, \qquad\qquad (5.7)$$

where

$$Y = [x^{T}(k + 1)\; x^{T}(k + 2)...x^{T}(k + P)]^{T}, \qquad\qquad (5.8)$$
$$U = [u(k - h)\; u(k - h + 1)...u(k - h + M - 1)]^{T}, \qquad\qquad (5.9)$$

$$G = \begin{pmatrix} I \\ F \\ \vdots \\ F^{P-1} \end{pmatrix}, \qquad\qquad (5.10)$$

$$A = \begin{pmatrix} B & 0 & 0 & \cdots & 0 \\ FB & B & 0 & \cdots & 0 \\ \vdots & \vdots & \ddots & & \vdots \\ F^{P-1}B & F^{P-2}B & F^{P-3}B & \cdots & F^{P-M}B \end{pmatrix}. \qquad\qquad (5.11)$$

Substituting (5.7) into (5.4), it follows that

$$J = (GFx(k) + AU)^{T}Q(GFx(k) + AU) + U^{T}RU, \qquad\qquad (5.12)$$

where

$$Q = diag\{Q(1), ..., Q(P)\} \geq 0,$$
$$R = diag\{R(1), ..., R(M)\} \geq 0.$$

The solution to the optimisation problem may be obtained from (5.12) by solving

$$\frac{\partial J}{\partial U} = 0.$$

It follows that the solution is

$$U = -(A^{T}QA + R)^{-1}A^{T}QGFx(k).$$

The current control is concerned with the first column. Thus, the result is

$$u(k - h) = -L_1(A^T QA + R)^{-1} A^T QGFx(k),$$
$$= -Dx(k), \tag{5.13}$$

where L_1 is a $1 \times n$ vector. Converting $u(k - h)$ in (5.13) back to $u(k)$, the GPC solution to the original system (5.3) with the index (5.4) is obtained as

$$u(k) = -Dx(k + h). \tag{5.14}$$

Note that once the prediction and control horizons with Q and R have been chosen, D is a constant matrix.

One problem left with (5.14) is to find the state signal $x(k + h)$ at time k. This is resolved by substituting (5.13) into (5.3) to obtain the closed-loop system,

$$x(k + 1) = (F - BD)x(k). \tag{5.15}$$

Thus, $x(k + h)$ can be expressed by the transmission of $x(k)$ as

$$x(k + h) = F^k x(k) \ \ 0 \le k < h, \tag{5.16}$$
$$x(k + h) = (F - BD)^h x(k) \ \ k \ge h. \tag{5.17}$$

Applying (5.17) or (5.16) to (5.14) produces the following result.

Theorem 5.1

For a controllable and observable system with time delay described by (5.3), the GPC control is given by:

$$u(k) = -L_1(A^T QA + R)^{-1} A^T QGF^{k+1} x(k) \ \ 0 \le k < h, \tag{5.18}$$
$$u(k) = -L_1(A^T QA + R)^{-1} A^T QGF[F$$
$$-BL_1(A^T QA + R)^{-1} A^T QGF]^h x(k) \ \ k \ge h. \tag{5.19}$$

Remark 5.1

From (5.19), it is seen that the control law given above includes the time-delay information and the current control $u(k)$ is actually a feedback of the future state at time $k + h$. This implies that the controller has a prediction capability in both the output and the control signal. Obviously, such design considerations can further improve the closed-loop performance.

5.2.2 Case of a Single-step Control Horizon

Consider a one-step control horizon, *i.e.*, $u(k - h + 1) = u(k - h + 2) = ... = u(k - h + M - 1) = 0$. This allows $(M - 1)$ steps of infinite weighting on the changes of control action ahead of one step of finite weight. This choice is valid as long as the system is stable with fairly damped poles as shown by Clarke *et al.* (1987). The resulting control law is naturally robust against unmodelled dynamics. Consequently, this version of GPC is particularly attractive for control applications.

Under the above conditions, the matrix A in (5.11) may be replaced by

$$A = \begin{pmatrix} B \\ FB \\ \vdots \\ F^{P-1}B \end{pmatrix} = A_1 B.$$

Thus, the control law given in (5.19) can be written as

$$\left.\begin{aligned} u(k) &= -(B^T A_1^T Q A_1 B + R)^{-1} B^T A_1^T Q G F^{k+1}, \quad 0 \le k \le h, \\ u(k) &= -(B^T A_1^T Q A_1 B + R)^{-1} B A_1^T Q G F [F \\ &\quad - B(B^T A_1^T Q A_1 B + R)^{-1} B^T A_1^T Q G F]^h x(k), \quad k \ge h. \end{aligned}\right\} \quad (5.20)$$

Using the result (5.13), the closed-loop system is given by

$$x(k + 1) = [I - B(B^T A_1^T Q A_1 + R)^{-1} B^T A_1^T Q G] F x(k). \quad (5.21)$$

5.2.3 Stability

In this sub-section, the general result on the stability of the system (5.3) with time delay is provided.

Theorem 5.2

The closed system (5.3) with the control law (5.14) or (5.20) is stable if all the eigenvalues of $[I - B(B^T A_1^T Q A_1 + R)^{-1} B^T A_1^T Q G] F$ are placed within the unit circle by an appropriate selection of the prediction horzion P and the weighting matrices Q and R.

Proof

The result follows directly from (5.15) and (5.21).

In the next section, the stability property will be discussed for an important class of systems commonly encountered in the system control industry, and an explicit form for which stability can be guaranteed will be provided.

5.2.4 PI Tuning for First-order Systems

In the system control industry, a large class of systems exhibit monotonic input–output transients with transfer functions which can be approximated (Luyben, 1990) as a first-order plus time-delay transfer function model described by

$$G(s) = \frac{b}{s + a} e^{-Ls}. \tag{5.22}$$

For this class of systems, it is adequate (Åström *et al.*, 1993) to consider a PI controller described by:

$$u(t) = k_p e(t) + k_i \int e(t) dt.$$

The use of derivative action is undesirable especially when the system has a long time delay associated with it. At the moment, tuning methods for PI(D) controllers are still primarily of trial and error types. Recently, there have been developments to associate PID tuning with shaping of the compensated frequency response (Åström *et al.*, 1993). However, these methods all operate on fixed control gains and may not be applicable when very tight and optimised control performance is required. This is especially true when faced with systems having more complex dynamics, such as those with non-minimum phase, time delay, or poorly damped and unstable modes. To this end, it will be very useful to incorporate the predictive and optimising features of the GPC developed in Section 5.2.2 into the simple PI control structure to further expand the capability of the conventional three-term controller. This will be done in the next section.

5.2.5 Controller Design Based on GPC Approach

Since the external set-point does not affect the feedback design, put $r = 0$. It then follows that $(s + a) = -be^{-Ls}u$, which is equivalent to the time-domain equation:

$$\dot{e} = -ae - bu(t - L). \tag{5.23}$$

Taking the z transform of (5.23), the following equation from the result of Warwick and Rees (1986) is obtained:

$$G(z) = \frac{\frac{b}{a}(1 - e^{-aT})z^{-h-1}}{1 - e^{-aT}z^{-1}}.$$

Thus, the discretised equation is

$$e(k + 1) = a'e(k) - b'u(k - h), \tag{5.24}$$

where $a' = e^{-aT}$ and $b' = \frac{b}{a}(1 - e^{-aT})$. The discrete PI controller is

$$u(k) = k_p e(k) + k_i \sum_{i=0}^{k-1} e(i). \tag{5.25}$$

Let $\theta(k) = \sum_{i=0}^{k-1} e(i)$. It then follows that

$$\theta(k + 1) = \theta(k) + e(k). \tag{5.26}$$

If the system states are chosen as $x = [\theta, e]^T$, an equivalent state equation to (5.24) is

$$x(k + 1) = \begin{pmatrix} 1 & 1 \\ 0 & a' \end{pmatrix} x(k) + \begin{pmatrix} 0 \\ -b' \end{pmatrix} u(k - h). \tag{5.27}$$

Clearly, both states are available and the state feedback of $Kx(k)$ is simply PI control, where $K = [k_p \;\; k_i]$. Thus, the result (5.20) obtained from the GPC solution of Section 5.2.2 may be applied to yield the PI parameters.

Comparing (5.25) with (5.20), the following results can be obtained:

$$(k_i \; k_p) = -(B^T A_1^T Q A_1 B + R)^{-1} B A_1^T Q G F^{k+1} \quad 0 \le k < h, \tag{5.28}$$

$$
\begin{aligned}
(k_i \; k_p) = &-(B^T A_1^T Q A_1 B + R)^{-1} B A_1^T Q G F [F \\
&- B(B^T A_1^T Q A_1 B + R)^{-1} B^T A_1^T Q G F]^h \quad k \ge h,
\end{aligned}
\tag{5.29}
$$

where

$$A_1 = \begin{pmatrix} 1 & 1 \\ 0 & a' \end{pmatrix}, B = \begin{pmatrix} 0 \\ -b' \end{pmatrix}.$$

Theorem 5.3 then follows directly from a simplification of (5.28) and (5.29).

Theorem 5.3

The GPC control for the system (5.24) with the state equation (5.27) may be posed in the form of an equivalent PI controller (5.25), where for $0 \leq k < h$,

$$k_i(k) = \frac{q_1 b' g_1}{q_1 b'^2 (g_2 + g_3 \bar{q}_2) + r}, \quad (\bar{q}_2 = \frac{q_2}{q_1} - 1),$$

$$k_p(k) = \frac{q_1 b'(g_1 \frac{a'^{k+1}-1}{a'-1} + g_2 a'^{k+1} + g_3 \bar{q}_2 a'^{k+1})}{q_1 b'^2 (g_2 + g_3 \bar{q}_2) + r},$$

and for $k \geq h$,

$$k_i(k) = \frac{q_1 b'[g_1 f_{11}(h) + (g_1 + g_2 a' + g_3 \bar{q}_2 a') f_{21}(h)]}{q_1 b'^2 (g_2 + g_3 \bar{q}_2) + r},$$

$$k_p(k) = \frac{q_1 b'[g_1 f_{12}(h) + (g_1 + g_2 a' + g_3 \bar{q}_2 a') f_{22}(h)]}{q_1 b'^2 (g_2 + g_3 \bar{q}_2) + r}.$$

g_1, g_2, and g_3 are constants as given in (5.32), $f_{ij}, i, j = 1, 2$, are given in (5.33), and q_1, q_2 and r are tuning parameters.

Proof

To obtain the feedback gains in (5.29) explicitly, one needs to calculate $(B^T A_1^T Q A_1 B + R)^{-1} B A_1^T Q G F^{k+1}$ and $[F - B(B^T A_1^T Q A_1 B + R)^{-1} B^T A_1^T \times Q G F]^h$.

Let $R = r$ and $Q(i) = \mathrm{diag}\{q_1, q_2\}$ where $i = 1, 2, ...P$. It can be derived that

$$A_1^T Q A_1 = \left[\begin{pmatrix} 1 & 0 \\ 0 & 1 \end{pmatrix} \begin{pmatrix} 1 & 0 \\ 1 & a' \end{pmatrix} \begin{pmatrix} 1 & 0 \\ 1+a' & a'^2 \end{pmatrix} \cdots \right.$$
$$\left. \begin{pmatrix} 1 & 0 \\ a'^{P-2} + a'^{P-3} + ... + 1 & a'^{P-1} \end{pmatrix} \right]$$

$$\times \begin{bmatrix} \begin{pmatrix} q_1 & 0 \\ 0 & q_2 \end{pmatrix} & & & \bigcirc \\ & \begin{pmatrix} q_1 & 0 \\ 0 & q_2 \end{pmatrix} & & \\ & & \begin{pmatrix} q_1 & 0 \\ 0 & q_2 \end{pmatrix} & \\ & & & \ddots \\ \bigcirc & & & \begin{pmatrix} q_1 & 0 \\ 0 & q_2 \end{pmatrix} \end{bmatrix}$$

$$
\times
\begin{bmatrix}
\begin{pmatrix} 1 & 0 \\ 0 & 1 \end{pmatrix} \\
\begin{pmatrix} 1 & 1 \\ 0 & a' \end{pmatrix} \\
\begin{pmatrix} 1 & 1+a' \\ 0 & a'^2 \end{pmatrix} \\
\vdots \\
\begin{pmatrix} 1 & a'^{P-2} + a'^{P-3} + \ldots + 1 \\ 0 & a'^{P-1} \end{pmatrix}
\end{bmatrix},
$$

$$
= \left[\begin{pmatrix} 1 & 0 \\ 0 & 1 \end{pmatrix} \begin{pmatrix} 1 & 0 \\ 1 & a' \end{pmatrix} \begin{pmatrix} 1 & 0 \\ 1+a' & a'^2 \end{pmatrix} \cdots \right.
$$
$$
\left. \begin{pmatrix} 1 & 0 \\ a'^{P-2} + a'^{P-3} + \ldots + 1 & a'^{P-1} \end{pmatrix} \right]
$$

$$
\times
\begin{bmatrix}
\begin{pmatrix} q_1 & 0 \\ 0 & q_2 \end{pmatrix} \times \begin{pmatrix} 1 & 0 \\ 0 & 1 \end{pmatrix} \\
\begin{pmatrix} q_1 & 0 \\ 0 & q_2 \end{pmatrix} \times \begin{pmatrix} 1 & 1 \\ 0 & a' \end{pmatrix} \\
\begin{pmatrix} q_1 & 0 \\ 0 & q_2 \end{pmatrix} \times \begin{pmatrix} 1 & 1+a' \\ 0 & a'^2 \end{pmatrix} \\
\vdots \\
\begin{pmatrix} q_1 & 0 \\ 0 & q_2 \end{pmatrix} \times \begin{pmatrix} 1 & a'^{P-2} + a'^{P-3} + \ldots + 1 \\ 0 & a'^{P-1} \end{pmatrix}
\end{bmatrix},
$$

$$
= q_1 \left[\begin{pmatrix} 1 & 0 \\ 0 & 1 \end{pmatrix} \begin{pmatrix} 1 & 0 \\ 1 & a' \end{pmatrix} \begin{pmatrix} 1 & 0 \\ 1+a' & a'^2 \end{pmatrix} \cdots \right.
$$
$$
\left. \begin{pmatrix} 1 & 0 \\ a'^{P-2} + a'^{P-3} + \ldots + 1 & a'^{P-1} \end{pmatrix} \right]
$$

$$
\times
\begin{bmatrix}
\begin{pmatrix} 1 & 0 \\ 0 & 1+\bar{q}_2 \end{pmatrix} \times \begin{pmatrix} 1 & 0 \\ 0 & 1 \end{pmatrix} \\
\begin{pmatrix} 1 & 0 \\ 0 & 1+\bar{q}_2 \end{pmatrix} \times \begin{pmatrix} 1 & 1 \\ 0 & a' \end{pmatrix} \\
\begin{pmatrix} 1 & 0 \\ 0 & 1+\bar{q}_2 \end{pmatrix} \times \begin{pmatrix} 1 & 1+a' \\ 0 & a'^2 \end{pmatrix} \\
\vdots \\
\begin{pmatrix} 1 & 0 \\ 0 & 1+\bar{q}_2 \end{pmatrix} \times \begin{pmatrix} 1 & a'^{P-2} + a'^{P-3} + \ldots + 1 \\ 0 & a'^{P-1} \end{pmatrix}
\end{bmatrix}
$$

(where $\bar{q}_2 = \dfrac{q_2}{q_1} - 1$),

$$
= q_1 \overbrace{\left[1 + (a' + 1) + \ldots + (a'^{P-2} + a'^{P-3} + \ldots + 1) \right.}^{P}
$$

$$\left.\begin{array}{c} 1 + (a' + 1) + ... + (a'^{P-2} + a'^{P-3} + ... + 1) \\ 1 + a'^2 + ... + (a'^{P-2} + a'^{P-3} + ... + 1)^2 + a'^{2(P-1)} \end{array}\right]$$

$$+ q_1 \begin{bmatrix} 0 & 0 \\ 0 & \bar{q}_2(1 + a'^2 + a'^4 + ... + a'^{2(P-1)}) \end{bmatrix},$$

$$= q_1 \begin{bmatrix} P & \frac{a'^P - a'}{(a'-1)^2} - \frac{P-1}{a'-1} \\ \frac{a'^P - a'}{(a'-1)^2} - \frac{P-1}{a'-1} & a_{qp} \end{bmatrix}, \tag{5.30}$$

where

$$a_{qp} = \frac{a'^{2P} - 2a'^{P+1} - 2a'^P - a'^4 + 2a'^3 + 2a'^2}{(a'^2 - 1)(a' - 1)^2}$$

$$+ \frac{a'^{2P} - 1}{a'^2 - 1} + \frac{P-2}{(a'-1)^2} + \frac{a'^{2P} - 1}{a'^2 - 1}\bar{q}_2. \tag{5.31}$$

Notice that $B^T A_1 Q G = B^T A_1 Q A_1$ and let

$$\left.\begin{array}{l} g_1 = \frac{\frac{a'^P - a'}{(a'-1)^2} - \frac{P-1}{(a'-1)}}{} \\ g_2 = \frac{a'^{2P} - 2a'^{P+1} - 2a'^P - a'^4 + 2a'^3 + 2a'^2}{(a'^2-1)(a'-1)^2} + \frac{a'^{2P}-1}{a^2-1} + \frac{P-2}{(a'-1)^2} \\ g_3 = \frac{a'^{2P} - 1}{a'^2 - 1} \end{array}\right\}. \tag{5.32}$$

Then, it follows that

$$B^T A_1 Q G = B^T A_1 Q A_1 = q_1[0 \quad -b'] \begin{bmatrix} P & \frac{a'^P - a'}{(a'-1)^2} - \frac{P-1}{a'-1} \\ g_1 & g_2 + g_3\bar{q}_2 \end{bmatrix},$$

$$= -q_1 b'[g_1 \quad g_2 + g_3\bar{q}_2].$$

For $B^T A_1 Q A_1 B + r$, this is calculated by

$$B^T A_1^T Q A_1 B + r = q_1 b'^2 (g_2 + g_3\bar{q}_2) + r.$$

Thus, it follows that

$$(B^T A_1^T Q A_1 B + R)^{-1} B^T A_1 Q G F^{k+1} = -\frac{q_1 b'[g_1 \quad g_2 + g_3\bar{q}_2] \begin{bmatrix} 1 & 1 \\ 0 & a' \end{bmatrix}^{k+1}}{q_1 b'^2 (g_2 + g_3\bar{q}_2) + r},$$

$$= -\frac{q_1 b'[g_1 \quad g_2 + g_3\bar{q}_2] \begin{bmatrix} 1 & \frac{a'^{k+1} - 1}{a'-1} \\ 0 & a'^{k+1} \end{bmatrix}}{q_1 b'^2 (g_2 + g_3\bar{q}_2) + r},$$

$$= -\frac{q_1 b'[g_1 \quad g_1 \frac{a'^{k+1}-1}{a'-1} + (g_2 + g_3\bar{q}_2)a'^{k+1}]}{q_1 b'^2 (g_2 + g_3\bar{q}_2) + r}.$$

Letting

$$[F - B(B^T A_1^T Q A_1 B + R)^{-1} B^T A_1^T Q G F]^h = \begin{bmatrix} f_{11}(h) & f_{12}(h) \\ f_{21}(h) & f_{22}(h) \end{bmatrix} \quad (5.33)$$

yields

$$(B^T A_1^T Q A_1 B + R)^{-1} B^T A_1 Q G F [F$$
$$-B(B^T A_1^T Q A_1 B + R)^{-1} B^T A_1^T Q G F]^h$$
$$= -\frac{q_1 b'[g_1 \quad g_1 + (g_2 + g_3\bar{q}_2)a'] \begin{bmatrix} f_{11}(h) & f_{12}(h) \\ f_{21}(h) & f_{22}(h) \end{bmatrix}}{q_1 b'2(g_2 + g_3\bar{q}_2) + r}$$
$$= -q_1 b'[g_1 f_{11}(h) + (g_1 + g_2 a' + g_3\bar{q}_2 a') f_{21}(h) g_1 f_{12}(h)$$
$$+(g_1 + g_2 a' + g_3\bar{q}_2 a') f_{22}(h)]/[q_1 b'2(g_2 + g_3\bar{q}_2) + r]$$

5.2.6 Selection of Weighting Matrices

The selection of weighting matrices Q and R has a direct and significant effect on the system performance. The overall performance of the control system may therefore be quite sensitive to the knowledge of the system operator or control engineer who may not be appreciative of the implications of these weighting matrices. On the other hand, classical control specifications such as the damping ratio and natural frequency are already well known and common concepts. To this end, it is useful to link the selection of weighting matrices to the desirable classical specifications, $i.e.$, a direct relationship between the weighting matrices Q and R, and classical specifications such as the damping ratio ξ and natural frequency ω_n of the closed-loop system. These relationships are provided in Theorem 5.4.

Theorem 5.4

When time $k > h$, the damping ratio ξ and natural frequency ω_n of the GPC-based closed-loop system in (5.27) and (5.25) are related to the weighting matrices q_1, q_2 and r by:

$$q_1 = \frac{q_{w\xi}a'r}{b'^2 g_1 e^{-2\xi w_n T}},$$

$$q_2 = \frac{(a' - e^{-2\xi w_n T})g_1 - q_{w\xi}a'(g_2 - g_3)}{b'^2 g_1 g_3 e^{-2\xi w_n T}} r,$$

where $q_{w\xi} = [1 - 2e^{-\xi w_n T}\cos(w_n\sqrt{1 - \xi^2}T) + e^{-2\xi w_n T}]$.

Proof

Since $(B^T A_1^T Q A_1 B + r)^{-1}$ is a scalar value, it follows that

$$B(B^T A_1^T Q A_1 B + r)^{-1} B^T A_1^T Q G = (B^T A_1^T Q A_1 B + r)^{-1} BB^T A_1^T Q G.$$

The next step is to calculate $BB^T A_1^T GF$ which yield,

$$BB^T A_1^T Q G = BB^T A_1^T Q A_1 = q_1 b'^2 \begin{bmatrix} 0 & 0 \\ g_1 & g_2 + g_3 \bar{q}_2 \end{bmatrix}.$$

The following result can be shown directly.

$$B(B^T A_1^T Q A_1 B + R)^{-1} B^T A_1^T Q G = (B^T A_1^T Q A_1 B + r)^{-1}$$
$$\times BB^T A_1^T Q A_1$$

$$= \frac{q_1 b'^2 \begin{bmatrix} 0 & 0 \\ g_1 & g_2 + g_3 \bar{q}_2 \end{bmatrix}}{q_1 b'^2 [g_2 + g_3 \bar{q}_2] + r}. \qquad (5.34)$$

Thus, it follows that

$$F - B(B^T A_1^T Q A_1 B + r)^{-1} B^T A_1^T Q G F$$
$$= \begin{bmatrix} 1 & 1 \\ -\frac{q_1 b'^2 g_1}{q_1 b'^2 (g_2 + g_3 \bar{q}_2) + r} & \frac{a'r - q_1 b'^2 g_1}{q_1 b'^2 (g_2 + g_3 \bar{q}_2) + r} \end{bmatrix}. \qquad (5.35)$$

Following (5.35),

$$x(k+1) = \left(-\frac{1}{\frac{q_1 b'^2 g_1}{q_1 b'^2 (g_2 + g_3 \bar{q}_2) + r}} \quad \frac{1}{\frac{a'r - q_1 b'^2 g_1}{q_1 b'^2 (g_2 + g_3 \bar{q}_2) + r}} \right) x(k). \qquad (5.36)$$

Applying z transformation to (5.36), the characteristic equation is obtained as

$$z^2 - \left(1 + \frac{a'r - q_1 b'^2 g_1}{q_1 b'^2 (g_2 + g_3 \bar{q}_2) + r}\right)z + \frac{a'r}{q_1 b'^2 (g_2 + g_3 \bar{q}_3) + r} = 0.$$

Comparing with the standard second-order characteristic equation (Jacquot, 1981),

$$z^2 - (e^{-\xi\omega_n T + j\omega_n \sqrt{1-\xi^2}T} + e^{-\xi\omega_n T - j\omega_n \sqrt{1-\xi^2}T})z + e^{-2\xi\omega_n T}, \qquad (5.37)$$

it follows that

$$1 + \frac{a'r - q_1 b'^2 g_1}{q_1 b'^2 (g_2 + g_3 \bar{q}_2) + r} = e^{-\xi\omega_n T + j\omega_n \sqrt{1-\xi^2}T}$$

$$+ e^{-\xi\omega_n T - j\omega_n \sqrt{1-\xi^2}T}, \qquad (5.38)$$

$$\frac{a'r}{q_1 b'^2 (g_2 + g_3 \bar{q}_2) + r} = e^{-2\xi\omega_n T}. \qquad (5.39)$$

Direct simplification of (5.38) and (5.39) yields the relationships contained in Theorem 5.4.

Remark 5.2

It may be noted that there are two equations in Theorem 5.4, but three unknown parameters q_1, q_2 and r in the weighting matrices. However, it is the relative magnitudes of these parameters that really matter in the control design. Therefore, the absolute value of r is not directly useful and significant in the design of controller gain in (5.25) and in practice, one can always fix $r = 1$. q_1 and q_2 may then be obtained directly from Theorem 5.4.

5.2.7 Stability

Consider next the stability of the proposed system. Since Theorem 5.2 is not directly applicable and useful, the stability problem of system (5.27) with control law (5.20) will be discussed. The end objective is to obtain an explicit condition to ensure closed-loop stability.

To prove the stability of the system (5.27) with time delay, it only needs to be proven that for this class of systems, the matrix of $F - B(B^T A_1^T Q A_1 B + R)^{-1} B^T A_1^T Q G F$ is stable. The eigenvalues of the closed-loop system (5.35) are given by

$$\lambda_{1,2} = \frac{q_1 b'^2 (g_2 + g_3 \bar{q}_2 - g_1) + ra' \pm g_4}{2[q_1 b'^2 (g_2 + g_3 \bar{q}_2) + r]},$$

where $g_4 = \sqrt{[q_1 b'^2 (g_2 + g_3 \bar{q}_2 - g_1) + ra']^2 - 4ra'[q_1 b'^2 (g_2 + g_3 \bar{q}_2) + r]}$. Theorem 5.5 then follows directly.

Theorem 5.5

For the system (5.27) with time delay, if the eigenvalues $|\lambda_1| < 1$ and $|\lambda_2| < 1$ with an appropriate choice of q_1, q_2 and r, then the controller given in (5.25),(5.29) stabilises the system.

Another general result on stability applicable to the system (5.27) is given in Theorem 5.6.

Theorem 5.6

For the system (5.27) with time delay, if $a' \neq 1$, $Q = \text{diag}\{Q(1), Q(2), ..., Q(P)\}$, where $Q(i) = diag\{q_1, q_2\}$, the prediction horizon is infinite, then the controller given in (5.25),(5.29) stabilises the system.

Proof

If $|a'| > 1$, (5.34) may be rewritten as

$$B(B^T A_1^T Q A_1 B + R)^{-1} B^T A_1^T G =$$

$$q_1 b'^2 \frac{\left[\begin{array}{cc} 0 & 0 \\ \frac{a'^{-P}-a'^{-2P+1}}{(a-1)^2} - \frac{P-1}{(a'-1)a'^{2P}} \, a_P + \frac{1-a'^{-2P}}{a^2-1} & + \frac{P-2}{(a'-1)^2 a'^{2P}} + \frac{1-a'^{-2P}}{a'^2-1} \bar{q}_2 \end{array} \right]}{q_1 b'^2 [a_P + \frac{1-a'^{-2P}}{a'^2-1} + \frac{P-2}{(a'-1)^2 a'^{2P}} + \frac{1-a'^{-2P}}{a'^2-1} \bar{q}_2] + \frac{r}{a'^{2p}}},$$

where

$$a_P = \frac{1 - 2a'^{-P+1} - 2a'^{-P} - a'^{-2P+4} + 2a'^{-2P+3} + 2a'^{-2P+2}}{(a'^2 - 1)(a' - 1)^2}.$$

Furthermore, it is noted that

$$\lim_{P \to \infty} \frac{P-1}{a'^{2P}} = 0, \tag{5.40}$$

$$\lim_{P \to \infty} \frac{P-2}{a'^{2P+1}} = 0, \tag{5.41}$$

$$\lim_{P \to \infty} \frac{r}{a'^{2P}} = 0. \tag{5.42}$$

Thus, it follows that

$$\lim_{P \to \infty} B(B^T A_1^T Q A_1 B + r)^{-1} B^T A_1^T Q G$$

$$
= \frac{q_1 b'^2 \begin{bmatrix} 0 & 0 \\ 0 & \frac{(a'^2 - 2a' + 2)}{(a' - 1)^2 (a'^2 - 1)} + \frac{1}{a'^2 - 1} \bar{q}_2 \end{bmatrix}}{q_1 b'^2 [\frac{(a'^2 - 2a' + 2)}{(a' - 1)^2 (a'^2 - 1)} + \frac{1}{a'^2 - 1} \bar{q}_2]},
$$

$$
= \begin{bmatrix} 0 & 0 \\ 0 & 1 \end{bmatrix}. \tag{5.43}
$$

Finally, the closed-loop system is given by

$$
F - B(B^T A_1^T Q A_1 B + r)^{-1} B^T A_1^T Q G F = \begin{bmatrix} 1 & 1 \\ 0 & 0 \end{bmatrix}, \quad P \longrightarrow \infty. \tag{5.44}
$$

It can be concluded that if $|a'| > 1$, then the closed-loop system is stable following Chen (1999).

If $|a'| < 1$, (5.34) can be rewritten as:

$$
B(B^T A_1^T Q A_1 B + R)^{-1} B^T A_1^T G
$$

$$
= \frac{q_1 b'^2 \begin{bmatrix} 0 & 0 \\ \frac{a'^P - a'}{P(a' - 1)^2} - \frac{1 - 1/P}{a' - 1} & a_{qp}/P \end{bmatrix}}{q_1 b'^2 a_{qp}/P + r/P},
$$

where a_{qp} is given in (5.31).

Since

$$
\lim_{P \longrightarrow \infty} \frac{a'^P - a}{P(a' - 1)^2} = 0,
$$

$$
\lim_{P \longrightarrow \infty} \frac{1 - 1/P}{a' - 1} = \frac{1}{a' - 1},
$$

$$
\lim_{P \longrightarrow \infty} \frac{a'^{2P} - 2a'^{P+1} - 2a'^P - a'^4 + 2a'^3 + 2a'^2}{P(a'^2 - 1)(a' - 1)^2} = 0,
$$

$$
\lim_{P \longrightarrow \infty} \frac{a'^{2P} - 1}{P(a'^2 - 1)} = 0,
$$

$$
\lim_{P \longrightarrow \infty} \frac{1 - 2/P}{(a' - 1)^2} = \frac{1}{(a' - 1)^2},
$$

$$
\lim_{P \longrightarrow \infty} \frac{r}{P} = 0,
$$

it follows that

$$
F - B(B^T A_1^T Q A_1 B + r)^{-1} B^T A_1^T Q G F = \begin{bmatrix} 1 & 1 \\ a' - 1 & a' - 1 \end{bmatrix},
$$

$$P \longrightarrow \infty. \qquad (5.45)$$

The eigenvalues of (5.45) are zero and a'. Thus, if $|a'| < 1$, then the system (5.27) with the control law (5.20) remains stable if the prediction horizon is infinite or can be approximated as an infinite one. Combining the stability result for the case of $|a'| > 1$, Theorem 5.6 follows.

Remark 5.3

It has been conjectured that the GPC can stabilise many systems with time delay (Clarke *et al.*, 1987). Theorem 5.6 now provides a clear theoretical result guaranteeing the stability of the class of commonly encountered systems (5.27). The condition that the prediction horizon P is infinite in Theorem 5.6, may be further reduced to P is large. One can even find the value of P.

Let $\sigma(\cdot)$ denote the largest singular value. If Λ is a Hermitian matrix, $\sigma(\Lambda) = max\{|\lambda|, \lambda \in \lambda(\Lambda)\}$.

Define

$$g_{11} = \frac{a'^{-P} - a'^{-2P+1}}{(a-1)^2} - \frac{P-1}{(a'-1)a'^{2P}},$$

$$g_{21} = \frac{1 - 2a'^{-P+1} - 2a'^{-P} - a'^{-2P+4} + 2a'^{-2P+3} + 2a'^{-2P+2}}{(a'^2 - 1)(a' - 1)^2}$$

$$+ \frac{1 - a'^{-2P}}{a^2 - 1} + \frac{P-2}{(a'-1)^2 a'^{2P}}$$

$$= \frac{1}{(a'^2 - 1)(a' - 1)^2} + \frac{1 - a'^{2P}}{a'^2 - 1} + g_{210},$$

$$g_{31} = \frac{1 - a'^{-2P}}{a'^2 - 1},$$

$$g_{41} = q_1 b'^2 [g_{21} + g_{31}\bar{q}_2] + \frac{r}{a'^{2P}}$$

$$= q_1 b'^2 (\frac{1}{(a'^2 - 1)(a' - 1)^2} + g_{210}) + g_{31}q_2 + \frac{r}{a'^{2P}},$$

$$\zeta_u = min\{\eta_1/2, \eta_2\}.$$

Theorem 5.7

For the open-loop unstable system (5.27) ($|a'| > 1$), if $r = 0$, and $P > max\{N_{u1}, N_{u2}, N_{u3}\}$ and is an even number , where N_{u1} is chosen such that

$$\frac{a'^{-2(N_{u1}+1)}}{a'^2 - 1} < \zeta_u,$$

N_{u2} such that $|g_{210}| < \zeta_u$ and N_{u3} such that

$$0 < g_{11} < \zeta_u,$$

then the controller given in (5.25),(5.29) stabilises the system (5.27).

Proof

(5.34) may be rewritten as

$$B(B^T A_1^T Q A_1 + R)^{-1} B^T A_1^T G = \frac{q_1 b'^2 \begin{bmatrix} 0 & 0 \\ g_{11} & g_{21} + g_{31}\bar{q}_2 \end{bmatrix}}{q_1 b'^2 [g_{21} + g_{31}\bar{q}_2] + \frac{r}{a'^2 P}}.$$

Thus, it follows that

$$F - B(B^T A_1^T Q A_1 B + R)^{-1} B^T A_1^T G F$$

$$= \begin{bmatrix} 1 & 1 \\ \frac{-q_1 b'^2 g_{11}}{q_1 b'^2 (g_{21} + g_{31}\bar{q}_2) + \bar{r}} & \frac{a'\bar{r} - q_1 b'^2 g_{11}}{q_1 b'^2 (g_{21} + g_{31}\bar{q}_2) + \bar{r}} \end{bmatrix}.$$

If $r = 0$, the eigenvalues of the closed-loop system are zero and

$$\frac{-q_1 b'^2 g_{11} + q_1 b'^2 (g_{21} + g_{31}\bar{q}_2)}{q_1 b'^2 (g_{21} + g_{31}\bar{q}_2)}. \tag{5.46}$$

For stability, it is essential to prove that the value of (5.46) is less than one. Since

$$\lim_{P \to \infty} \frac{a'^{-2P}}{a'^2 - 1} = 0,$$

this implies that there exists an N_{u1} such that for a given $\zeta_u > 0$, when $P > N_{u1}$, $-\zeta_u < \frac{a'^{-2P}}{a'^2 - 1} < \zeta_u$. Similarly,

$$\lim_{P \to \infty} g_{210} = 0,$$

which implies that there exists an N_{u2} such that for a given $\zeta_u > 0$, when $P > N_{u2}$, $-\zeta_u < g_{210} < \zeta_u$;

$$\lim_{P \to \infty} g_{11} = 0,$$

and

$$g_{11} = \frac{a'^P - (Pa' + 1 - P)}{(a' - 1)^2 a'^{2P}},$$

which imply that there exists an N_{u3} such that for a given $\zeta_u > 0$, when $P > N_{u3}$ and is an even number, $0 < g_{11} < \zeta_u$.

Since $|a'| > 1$, one can infer that $\frac{1}{(a'^2-1)(a'-1)^2} = \eta_1 > 0$ and $\frac{1}{a'^2-1} = \eta_2 > 0$. Let $\zeta_u = min\{\eta_1/2, \eta_2\}$. Thus, when $P > max\{N_{u1}, N_{u2}, N_{u3}\}$, it follows that

$$q_1 b'^2 [\frac{1}{(a'^2 - 1)(a' - 1)^2} + g_{210}] + q_2 b'^2 g_{31} > 0,$$

and

$$q_1 b'^2 [\frac{1}{(a'^2 - 1)(a' - 1)^2} + g_{210}] + q_2 b'^2 g_{31} > q_1 b'^2 [\frac{1}{(a'^2 - 1)(a' - 1)^2}$$
$$+ g_{210}] + q_2 b'^2 g_{31}$$
$$- q_1 b'^2 g_{11} > 0.$$

Then, the eigenvalue given in (5.46) is within the unit circle. The theorem is proven.

For the open-loop stable system, *i.e.*, $|a'| < 1$, define

$$g_{12} = \frac{a'^P - a'}{P(a' - 1)} + \frac{1}{P},$$

$$g_{22} = \frac{a'^{2P} - 2a'^{P+1} - 2a'^P - a'^4 + 2a'^3 + 2a'^2}{P(a' + 1)(a' - 1)^2} + \frac{2}{P(1 - a')},$$

$$g_{32} = \frac{a'^{2P} - 1}{P(a' - 1)},$$

$$g_{42} = q_1 b'^2 g_{22} + q_2 b'^2 g_{32} + r/P,$$

$$g_{52} = q_1 b'^2 (\frac{1}{a' - 1} + g_{22} + g_{32} q_2/q_1) + r/P,$$

$$\eta_3 = \frac{1}{|a' - 1|} q_1 b'^2 - \frac{|a'|}{|a' - 1|} q_1 b'^2.$$

Then, (5.34) can be rewritten as:

$$B(B^T A_1^T Q A_1 B + R)^{-1} B^T A_1^T G = \frac{q_1 b'^2 \begin{bmatrix} 0 & 0 \\ g_{12} - 1 \frac{1}{a'-1} + g_{22} + g_{32}q_2/q_1 \end{bmatrix}}{q_1 b'^2 (\frac{1}{a'-1} + g_{22} + g_{32}q_2/q_1) + r/P}.$$

Therefore, it follows that

$$F - B(B^T A_1^T Q A_1 B + R)^{-1} B^T A_1 Q G F$$

$$= \begin{bmatrix} 1 & 1 \\ \frac{q_1 b'^2 (1-g_{12})}{q_1 b'^2 (\frac{1}{a'-1}+g_{22}+g_{32}q_2/q_1)+r/P} & \frac{q_1 b'^2 (1-g_{12}+a'r/P)}{q_1 b'^2 (\frac{1}{a'-1}+g_{22}+g_{32}q_2/q_1)+r/P} \end{bmatrix}$$

$$= \begin{bmatrix} 1 & 1 \\ \frac{q_1 b'^2 (1-g_{12})}{q_1 b'^2 \frac{1}{a-1}+g_{42}} & \frac{q_1 b'^2 (1-g_{12}+a'r/P)}{q_1 b'^2 \frac{1}{a'-1}+g_{42}} \end{bmatrix}$$

$$= \begin{bmatrix} 1 & 1 \\ \frac{q_1 b'^2}{\frac{1}{a'-1}q_1 b'^2+g_{42}} & \frac{q_1 b'^2}{\frac{1}{a'-1}q_1 b'^2+g_{42}} \end{bmatrix}$$

$$+ g_{12}q_1/g_{52} \begin{bmatrix} 0 & 0 \\ -b'^2 & -b'^2 \end{bmatrix} + \bar{r}_2/g_{52} \begin{bmatrix} 0 & 0 \\ 0 & a' \end{bmatrix}$$

$$= \Lambda + k_1 E_{21} + k_2 E_{22},$$

where

$$\Lambda = \begin{bmatrix} 1 & 1 \\ \frac{q_1 b'^2}{\frac{1}{a'-1}q_1 b'^2+g_{42}} & \frac{q_1 b'^2}{\frac{1}{a'-1}q_1 b'^2+g_{42}} \end{bmatrix},$$

$$= \begin{bmatrix} 1 & 1 \\ g & g \end{bmatrix}, \tag{5.47}$$

$$\bar{r}_2 = r/P, \tag{5.48}$$

$$k_1 = g_{12}q_1/g_{52}, \tag{5.49}$$

$$k_2 = \bar{r}_2/g_{52}. \tag{5.50}$$

Now the problem is to analyse the stability of the following system:

$$Z(k+1) = \Lambda Z(k) + \sum_i k_i E_{2i}, \quad i = 1, 2. \tag{5.51}$$

If the matrix Λ is stable, the solution of the following Lyapunov equation:

$$\Lambda^T M \Lambda - M = -2I,$$

always exists and produces

$$M = \begin{bmatrix} m_1 \ m_2 \\ m_2 \ m_1 \end{bmatrix}$$

$$= m_1 \begin{bmatrix} 1 \ 0 \\ 0 \ 1 \end{bmatrix} + m_2 \begin{bmatrix} 0 \ 1 \\ 1 \ 0 \end{bmatrix}$$

$$= \sum_i m_i E_{Mi}, \ i = 1, 2,$$

where

$$m_1 = \frac{-2 + 4g + 4g^2}{g^2 + 2g}, m_2 = \frac{-2 + 2g^2}{2g + g^2}.$$

The matrix Λ can also be expressed as:

$$\Lambda = \begin{bmatrix} 1 \ 1 \\ g \ g \end{bmatrix} = \begin{bmatrix} 1 \ 1 \\ 0 \ 0 \end{bmatrix} + g \begin{bmatrix} 0 \ 0 \\ 1 \ 1 \end{bmatrix}$$

$$= \sum_i g_{\Lambda i} E_{\Lambda i}, \ i = 1, 2,$$

where $g_{\Lambda 1} = 1$.

Define

$$M_{\Lambda mijk} = \frac{1}{2}(E_{\Lambda j}^T E_{mk} E_{2i} + E_{2i}^T E_{mk} E_{\Lambda j}),$$

$$M_{mijk} = \frac{1}{2}(E_{2i}^T E_{mk} E_{2j} + E_{2j}^T E_{mk} E_{2i}),$$

$$M_{mij} = \frac{1}{2} E_{2i}^T E_{mj} E_{2i}.$$

The following theorem shows that the system (5.27) ($|a'| < 1$) can be stabilised by the controller (5.25),(5.29).

Theorem 5.8

For the open-loop stable system (5.27), if $P > N_{s1}$ where N_{s1} is chosen such that

$$|g_{42}| < \eta_3/2$$

and

$$\rho^2 \left(\sum_{j,k} g_{\Lambda j}^2 m_k^2 + 2 \sum_k m_k^2 \right)$$

$$< 1/[\sum_{i,j,k} \sigma_{max}^2(M_{Amijk}) + \sum_{i,j,k} \sigma_{max}^2(M_{mijk}) + \sum_{i,j} \sigma_{max}^2(M_{mij})] \quad (5.52)$$

where $\rho = max\{k_1, k_2\}$ and is less than one, then the controller given in (5.25),(5.29) stabilises the system (5.27).

Proof

The eigenvalues of the matrix (5.47) are zero and

$$\frac{\frac{a'}{a'-1}q_1 b'^2 + g_{42}}{\frac{1}{a'-1}q_1 b'^2 + g_{42}}. \quad (5.53)$$

Note that

$$\lim_{P\to\infty} g_{42} = \lim_{P\to\infty} (q_1 b'^2 g_{22} + q_2 b'^2 g_{32} + \bar{r}_2) = 0. \quad (5.54)$$

Thus, there exists an N_{s1} such that for a given ζ_s, when $P > N_{s1}$,

$$|g_{42}| < \zeta_s. \quad (5.55)$$

Since $|a'| < 1$, it is noted that $\frac{1}{|a'-1|}q_1 b'^2 > \frac{|a'|}{|a'-1|}q_1 b'^2$ and $\eta_3 > 0$. Let $\zeta_s = \eta_3/2$. It is deduced that

$$|\frac{q_1 b'^2}{a'-1} + g_{42}| \geq |\frac{q_1 b^2}{a'-1}| - |g_{42}|, \quad (5.56)$$

and (5.55) holds for large P. It can be shown that

$$|\frac{q_1 b^2}{a'-1}| - |g_{42}| > |\frac{q_1 b^2}{a'-1}| - \zeta_s$$
$$= |\frac{q_1 b^2}{a'-1}| - \frac{1}{2}(\frac{1}{|a'-1|}q_1 b'^2 - \frac{|a'|}{|a'-1|}q_1 b'^2)$$
$$= \frac{1}{2}(\frac{1}{|a'-1|}q_1 b'^2 + \frac{|a'|}{|a'-1|}q_1 b'^2). \quad (5.57)$$

On the other hand,

$$|\frac{a'}{a'-1}q_1 b'^2 + g_{42}| \leq |\frac{a'}{a'-1}q_1 b'^2| + |g_{42}|$$

$$< \frac{|a'|}{|a'-1|} q_1 b'^2 + \eta_3$$

$$= \frac{|a'|}{|a'-1|} q_1 b'^2 + \frac{1}{2}(\frac{1}{|a'-1|} q_1 b'^2 - \frac{|a'|}{|a'-1|} q_1 b'^2)$$

$$= \frac{1}{2}(\frac{1}{|a'-1|} q_1 b'^2 + \frac{|a'|}{|a'-1|} q_1 b'^2). \tag{5.58}$$

Comparing (5.58) with (5.57), it follows that

$$|\frac{\frac{a'}{a'-1} q_1 b'^2 + g_{42}}{\frac{1}{a'-1} q_1 b'^2 + g_{42}}| < 1.$$

Thus, the matrix (5.47) is stable. Let the Lyapunov function be

$$v(Z) = Z^T M Z.$$

It can be shown that

$$\Delta v(Z) = Z(k+1)^T M Z(k+1) - Z(k)^T M Z(k)$$
$$= Z^T(k)(\Lambda^T + \sum_i k_i E_{2i}^T) M (\Lambda + \sum_i k_i E_{2i}) Z(k)$$
$$- Z^T(k) M Z(k)$$
$$= Z^T(k)[\Lambda^T M \Lambda - M + \sum_i k_i (\Lambda^T M E_{2i} + E_{2i}^T M \Lambda)$$
$$+ \sum_{ij} k_i k_j (E_{2i}^T M E_{2j} + E_{2j}^T M E_{2i}) + \sum_i k_i^2 E_{2i}^T M E_{2i}] Z(k)$$
$$= Z^T(k)(-2I + 2 \sum_{i,j,k} k_i g_{\Lambda j} m_k M_{\Lambda m i j k} + 2 \sum_{i,j,k} k_i k_j m_k M_{m i j k}$$
$$+ 2 \sum_{i,j} k_i^2 m_j M_{m i j}) Z(k).$$

Let

$$\Omega = -2I + 2 \sum_{i,j,k} k_i g_{\Lambda j} m_k M_{\Lambda m i j k} + 2 \sum_{i,j,k} k_i k_j m_k M_{m i j k}$$
$$+ 2 \sum_{i,j} k_i^2 m_j M_{m i j}.$$

For the system in (5.51) to be asymptotically stable, Δv is negative definite, or, equivalently, all the eigenvalues of Ω are negative; *i.e.*,

$$\lambda(\Omega) < 0. \tag{5.59}$$

Further,

$$\lambda(\Omega) \leq 2 \sum_{i,j,k} \lambda_{max}(k_i g_{\Lambda j} m_k M_{\Lambda mijk}) + 2 \sum_{i,j,k} \lambda_{max}(k_i k_j m_k M_{mijk})$$

$$+ 2 \sum_{i,j} k_i^2 m_j M_{mij} - 2$$

$$\leq 2 \sum_{i,j,k} |k_i| |g_{\Lambda j}| |m_k| \sigma_{max}(M_{\Lambda mijk})$$

$$+ 2 \sum_{i,j,k} |k_i| |k_j| |m_k| \sigma_{max}(M_{mijk})$$

$$+ 2 \sum_{i,j} |k_i^2| |m_j| \sigma_{max}(M_{mij}) - 2.$$

Thus, the system (5.51) is stable if

$$\sum_{i,j,k} |k_i| |g_{\Lambda j}| |m_k| \sigma_{max}(M_{\Lambda mijk}) + \sum_{i,j,k} |k_i| |k_j| |m_k| \sigma_{max}(M_{mijk})$$

$$+ \sum_{i,j} |k_i^2| |m_j| \sigma_{max}(M_{mij}) \leq 1. \tag{5.60}$$

Let $|k_i| < 1$. Thus, it follows that

$$\sum_{i,j,k} |k_i| |g_{\Lambda j}| |m_k| \sigma_{max}(M_{\Lambda mijk}) + \sum_{i,j,k} |k_i| |m_k| \sigma_{max}(M_{mijk})$$

$$+ \sum_{i,j} |k_i| |m_j| \sigma_{max}(M_{mij})$$

$$< \sum_{i,j,k} |k_i| |g_{\Lambda j}| |m_k| \sigma_{max}(M_{\Lambda mijk}) + \sum_{i,j,k} |k_i| |m_k| \sigma_{max}(M_{mijk})$$

$$+ \sum_{i,j} |k_i| |m_j| \sigma_{max}(M_{mij})$$

$$< \rho [\sum_{i,j,k} |g_{\Lambda j}| |m_k| \sigma_{max}(M_{\Lambda mijk}) + \sum_{i,j,k} |m_k| \sigma_{max}(M_{mijk})$$

$$+ \sum_{i,j} |m_j| \sigma_{max}(M_{mij})]$$

$$< \quad \rho(|g_{\Lambda 1}||m_1| \ |g_{\Lambda 1}||m_2|...|m_2|) \begin{pmatrix} \sigma_{max}(M_{\Lambda m111}) \\ \sigma_{max}(M_{\Lambda m112}) \\ \vdots \\ \sigma_{max}(M_{m22}) \end{pmatrix} < 1. \quad (5.61)$$

Thus, if (5.61) holds, then (5.60) is satisfied. Furthermore, if the following condition

$$\rho^2(\sum_{j,k} g_{\Lambda j}^2 m_k^2 + 2\sum_k m_k^2)[\sum_{i,j,k} \sigma_{max}^2(M_{\Lambda mijk}) + \sum_{i,j,k} \sigma_{max}^2(M_{mijk})$$
$$+ \sum_{i,j} \sigma_{max}^2(M_{mij})] < 1$$

is satisfied, then (5.61) holds. Therefore, if the condition (5.52) is satisfied, then (5.60) holds. Then, the theorem is established.

Remark 5.4

The conditions in Theorem 5.7 always hold when P is large. The conditions in Theorem 5.8 also always hold. When P is large, the right side of the condition (5.52) always holds since $M_{\Lambda ijk}, M_{mijk}, M_{mij}$ are constant matrices. Note that the right side of the condition (5.52)

$$\lim_{P\to\infty} \rho^2(\sum_{j,k} g_{\Lambda j}^2 m_k^2 + 2\sum_k m_k^2) = 0.$$

This implies that for a given value as shown in the right side of the condition (5.52), there exists an N such that when $P > N$, the right side of the condition (5.52) will always be less than the left side of (5.52). To obtain such a P, an initial value of N is needed to check if the condition (5.52) holds. If so, then $P = N$; otherwise, $P = N + 1$ and continue to check the condition until it is satisfied.

5.2.8 Robustness Analysis

For any model-based control system to be practically useful, it has to be robust to modelling perturbations since all models are inaccurate in practice. In this section, the robustness properties of the control system will be investigated.

Consider the stability of the controlled system in (5.3) and (5.20) with the real parameters perturbed to F_r, B_r and h_r. Since the system signals cannot go to infinity over a finite time interval, the robustness properties are considered when time $k \geq \max\{h, h_r\}$.

When the control law (5.20) is applied to the system (5.27), the resulting closed-loop system is

$$x(k+1) = F_r x(k) - B_r D F_c^h x(k - h_r),$$ (5.62)

where $F_c = F - BD$. Let matrix $\hat{F}_c$ be the solution of equation:

$$\hat{F}_c = F_r - B_r D F_c^h \hat{F}_c^{-h_r},$$ (5.63)

then the following theorem is established.

Theorem 5.9

The perturbed system with real system parameters F_r, B_r and h_r is stable if $\hat{F}_c$ given by (5.63) is stable.

Proof

Substituting (5.63) into (5.62) yields:

$$x(k+1) = (\hat{F}_c + B_r D F_c^h \hat{F}_c^{-h_r}) x(k) - B_r D F_c^h x(k - h_r).$$ (5.64)

Consider the equation:

$$x(k+1) = \hat{F}_c x(k).$$

It follows that

$$x(k - h_r) = \hat{F}_c^{-h_r} x(k).$$ (5.65)

Applying (5.64) to (5.65), it follows that

$$x(k+1) = \hat{F}_c x(k).$$ (5.66)

Thus, if (5.66) is stable, then the stability of (5.62) is ensured.

Now a special case of the PI controller (5.25) will be analysed for the system (5.27) with real parameters

$$F_r = F + \delta_F \begin{pmatrix} 0 & 0 \\ 0 & 1 \end{pmatrix},$$
$$= F + \delta_F E_F, \tag{5.67}$$
$$B_r = B + \delta_B \begin{pmatrix} 0 \\ 1 \end{pmatrix},$$
$$= B + \delta_B E_B, \tag{5.68}$$

and h_r. Let

$$F_c^h \hat{F}_c^{-h_r} = I + \Delta = I + \delta_{F_{c1}} \begin{pmatrix} 1 & 0 \\ 0 & 0 \end{pmatrix} + \delta_{F_{c2}} \begin{pmatrix} 0 & 1 \\ 0 & 0 \end{pmatrix}$$
$$+ \delta_{F_{c3}} \begin{pmatrix} 0 & 0 \\ 1 & 0 \end{pmatrix} + \delta_{F_{c4}} \begin{pmatrix} 0 & 0 \\ 0 & 1 \end{pmatrix},$$
$$= I + \sum_{i=1}^{4} \delta_{F_{ci}} E_{F_{ci}}. \tag{5.69}$$

Substituting (5.67)–(5.69) into (5.63) yields

$$\hat{F}_c = (F + \delta_F E_F) - (B + \delta_B E_B)D(I + \sum_{i=1}^{4} \delta_{F_{ci}} E_{F_{ci}}),$$
$$= F - BD + \Delta \hat{F}_c, \tag{5.70}$$

where

$$\Delta \hat{F}_c = \delta_F E_F - \delta_B E_B D - BD \sum_{i=1}^{4} \delta_{F_{ci}} E_{F_{ci}}$$
$$- \sum_{i=1}^{4} \delta_B \delta_{F_{ci}} E_B D E_{F_{ci}}. \tag{5.71}$$

To determine the system robustness, it is necessary to introduce the following equation:

$$(F - BD)^T P(F - BD) - P = -2I, \tag{5.72}$$

where I is the unit matrix and P is the solution of (5.72).

Theorem 5.10 provides for a sufficient stability condition when the system perturbation is sufficiently small.

Theorem 5.10

Assume that the nominal system (5.27) is stabilised by Theorem 5.5. If the system parameter perturbations are sufficiently small, the control system in (5.27) with the control law (5.25) is robustly stable.

Proof

Since the asymptotic stability of $A - BD$ can be guaranteed by Theorem 5.5, it is well known that for a given matrix $2I > 0$, the solution P of (5.72) is positive definite. Hence, $V(x(k)) = x^T(k)Px(k)$ is a positive definite function. Then, it can be shown that

$$
\begin{aligned}
\Delta V(x(k) &= x^T(k+1)Px(k+2) - x(k)^T Px(k), \\
&= x^T(k)[(F - BD)^T P(F - BD) - P + (F - BD)^T P\Delta\hat{F}_c \\
&\quad + \Delta\hat{F}_c^T P(F - BD) + \Delta\hat{F}_c^T P\Delta\hat{F}_c]x(k), \\
&= x^T(k)[-2I + (F - BD)^T P\Delta\hat{F}_c + \Delta\hat{F}_c^T P(F - BD) \\
&\quad + \Delta\hat{F}_c^T P\Delta\hat{F}_c]x(k).
\end{aligned}
$$

Let

$$
M = -2I + (F - BD)^T P\Delta\hat{F}_c + \Delta\hat{F}_c^T P(F - BD) + \Delta\hat{F}_c^T P\Delta\hat{F}_c.
$$

For the system in (5.27),(5.25) to be asymptotically stable, $\Delta V(x)$ is negative definite, or equivalently, all the eigenvalues of M are negative, $i.e.,$

$$
\lambda(M) < 0. \tag{5.73}
$$

Furthermore,

$$
\begin{aligned}
\lambda_{max}(M) \leq -2 + \lambda_{max}[(F - BD)^T P\Delta\hat{F}_c + \Delta\hat{F}_c^T P(F - BD) \\
+ \Delta\hat{F}_c^T P\Delta\hat{F}_c].
\end{aligned}
$$

Thus, if

$$
\lambda_{max}[(F - BD)^T P\Delta\hat{F}_c + \Delta\hat{F}_c^T P(F - BD) + \Delta\hat{F}_c^T P\Delta\hat{F}_c] < 2,
$$

then (5.73) holds. It is noted that

$$
\begin{aligned}
\lambda_{max}[(F - BD)^T P\Delta\hat{F}_c + \Delta\hat{F}_c^T P(F - BD) + \Delta\hat{F}_c^T P\Delta\hat{F}_c] \\
\leq 2||F - BD||||P||||\Delta\hat{F}_c|| + ||\Delta\hat{F}_c||^2||P||.
\end{aligned}
$$

Assume $||\Delta \hat{F}_c|| < \delta (\delta < 1)$, then it is easy to show that

$$2||F - BD||||P||||\Delta \hat{F}_c|| + ||\Delta \hat{F}_c||^2||P|| < \delta||P||(2||F - BD|| + 1).$$

Therefore, the condition (5.73) can be satisfied if

$$\delta < \frac{1}{||P||(||F - BD|| + \frac{1}{2})}.$$

Let $|\delta_F| < \delta_0, |\delta_B| < \delta_0$, and $|\delta_{Fci}| < \delta_0$ $(\delta_0 < 1)$. Now

$$||\Delta \hat{F}_c|| \le |\delta_F|||E_F|| + |\delta_B|||E_B D|| + \sum_{i=1}^{4} |\delta_B||\delta_{Fci}|||E_B D E_{Fci}||$$

$$< \delta_0(||E_F|| + ||E_B D|| + \sum_{i=1}^{4} ||BDE_{Fci}|| + \sum_{i=1}^{4} ||E_B D E_{Fci}||). \quad (5.74)$$

In view of (5.74),

$$\delta_0 < \frac{\delta}{||E_F|| + ||E_B D|| + \sum_{i=1}^{4} ||BDE_{Fci}|| + \sum_{i=1}^{4} ||E_B D E_{Fci}||}. \quad (5.75)$$

Thus, the system (5.70) with the perturbation parameters is stable if

$$\delta_0 < min\{1, \frac{\delta}{||E_F||+||E_B D||+\sum_{i=1}^{4} ||BDE_{Fci}||+\sum_{i=1}^{4} ||E_B D E_{Fci}||}\} \quad (5.76)$$

Theorem 5.10 is thus proven.

5.2.9 On-line Tuning of the Control Weightings

To deal with variation of the system parameters, an on-line tuning method for adjusting q_1 and q_2 can be considered. The objective for this task is

$$J = \frac{1}{2}e^2(k). \quad (5.77)$$

The gradient search method is used to obtain on-line values of q_1 and q_2. Assume that q_1, q_2 will not change during each time-delay period. From (5.36), it can be seen that

$$q_1^{k+1} = q_1^k - \beta_1 \frac{\partial J}{\partial q_1}, \tag{5.78}$$

$$q_2^{k+1} = q_1^k - \beta_2 \frac{\partial J}{\partial q_2}, \tag{5.79}$$

where

$$\frac{\partial J}{\partial q_1} = e(k) \frac{\partial e(k)}{\partial q_1}$$

$$= -\frac{b'^4 g_1 g_3 (q_2 - q_1) + b'^2 r g_1}{[q_1 b'^2 (g_2 - g_3) + q_2 b'^2 g_3 + r]^2} \sum_{i}^{k-2} e(i) e(k)$$

$$- \frac{b'^4 g_1 g_3 (q_2 - q_1) + r b'^2 g_1 + a' r b'^2 g_2}{[q_1 b'^2 (g_2 - g_3) + q_2 b'^2 g_3 + r]^2} e(k-1) e(k),$$

$$\frac{\partial J}{\partial q_2} = e(k) \frac{\partial e(k)}{\partial q_2}$$

$$= -\frac{q_1 b'^4 g_1 g_3}{[q_1 b'^2 (g_2 - g_3) + q_2 b'^2 g_3 + r]^2} \sum_{i}^{k-2} e(i) e(k)$$

$$- \frac{a' b'^2 r g_3 - q_1 b'^4 g_1 g_3}{[q_1 b'^2 (g_2 - g_3) + q_2 b'^2 g_3 + r]^2} e(k-1) e(k).$$

5.2.10 Simulation Examples

To illustrate the effectiveness of the PI control system, simulation is studied with different types of common systems. In order to use the tuning method, the system has to be modelled as a first-order transfer function with time delay. There are numerous methods for first-order modelling. One may choose to apply least squares fitting method in either the time or frequency domain, or use the relay feedback test (Luyben, 1990; Åström and Hägglund, 1995) which is amenable to implementation on an automated platform.

In the simulation study, all the continuous transfer functions used will be discretised by using the function c2d of the MATLAB toolbox with a sampling interval of 0.005s for Example 5.1 and 0.01s for Examples 5.2 and 5.3. The prediction horizon will be chosen to be 10 steps. From extensive simulation and experiments, the choice of $\xi \in [0.7, 1.0]$ and $\frac{\omega h}{T} \in [1.0, 1.6]$ appears to be most satisfactory. Following these empirical guidelines, select $\xi = 1$ and $\frac{\omega_n h}{T} = 1.44$ for the simulation study. Finer tuning may of course be used to yield better performance.

Example 5.1

The example in Chapter 4 regarding injection moulding machine control is used again in this simulation.

To apply the control method, the fourth-order model in (4.59) must be approximated as a first-order transfer function with time delay which is computed from a least squares fitting as

$$G(s) = \frac{110.6315}{s + 105.2216} e^{-0.0017s}. \tag{5.80}$$

The discrete-system transfer function model with sampling interval 0.005s for the continuous time model is

$$G_P(z) = \frac{0.3941z^{-1} - 0.0085z^{-2} - 0.0073z^{-3} + 0.0011z^{-4}}{1 - 0.7805z^{-1} + 0.1538z^{-2} - 0.0121z^{-3}}. \tag{5.81}$$

In addition, Wang (1984), Agrawal, Pandelidis and Pecht (1987) indicated the need to add a pure delay in the input to physically reflect the delay associated with the electromechanical servovalves to complete transmitting the oil flow signal. For the particular machine of concern with the model (5.80), the time delay is about 25ms. Thus, $h = 25$ms is introduced into the model. The PI parameters for the controller are computed to be

$$\left. \begin{aligned} k_i(k) &= 25.6389 \\ k_p(k) &= 62.595 - 4.3023 \times 0.5904^k \end{aligned} \right\} \ 0 \leq k < h,$$

$$\left. \begin{aligned} k_i(k) &= 16.6015 \\ k_p(k) &= 6.444 \end{aligned} \right\} \ k \geq h.$$

The solid line in Fig. 5.2 shows that the system yields good set-point tracking using the PI controller. For the purpose of comparison, the PID injection moulding controller suggested by Pandelidis and Agrawal (1988) without considering the effect of time delay is commissioned to follow the same set-point. The control law is described by

$$u(k) = 0.45e(k) + 0.2331e(k-1) - 0.3057e(k-2) + u(k-1). \tag{5.82}$$

The dotted line in Fig. 5.2 shows the set-point response of the system under this control. The results show large variations in set-point tracking although the control performance is reasonable for the same system without time delay, as shown in Fig. 5.3.

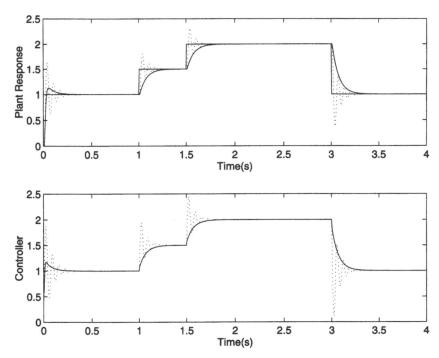

Fig. 5.2. Control of System 5.80

Example 5.2

Consider the high-order system

$$G_p(s) = \frac{1}{(s+1)^{10}}. \tag{5.83}$$

The first-order model for the controller is given by

$$\frac{1}{1+2.72s}e^{-7.69s}. \tag{5.84}$$

For comparison, PI tuning as defined by Smith and Corripio (1985) is employed. The PI parameters for the controller are computed to be

$$\left.\begin{array}{l} k_i(k) = 0.0003, \\ k_p(k) = 0.081 - 0.0774 \times 0.9963^k \end{array}\right\} \quad 0 \le k < h,$$

$$\left.\begin{array}{l} k_i(k) = 0.0638, \\ k_p(k) = 0.000199 \end{array}\right\} \quad k \ge h,$$

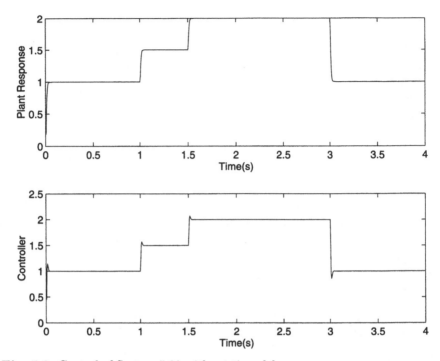

Fig. 5.3. Control of System 5.80 without time delay

while the control parameters according to Smith and Corripio (1985) are

$$k_i(k) = 0.1167.$$
$$k_p(k) = 0.00042913.$$

The closed-loop response in Fig. 5.4 shows that the optimal PI controller produces a much better performance than the Smith and Corripio (1985) method.

Example 5.3

Consider a non-minimum phase system

$$G_p(s) = \frac{1-s}{(1+s)^2(s+2)}. \tag{5.85}$$

The first-order model used to represent the system is

$$G(s) = \frac{0.4472}{s+0.8944}e^{-2.38s}. \tag{5.86}$$

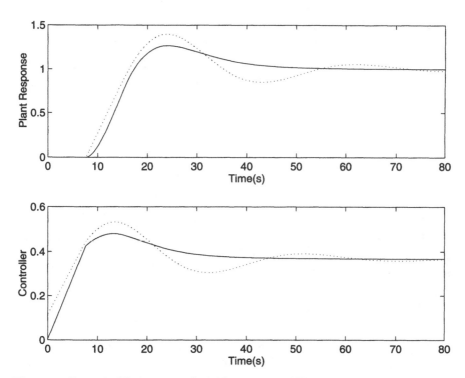

Fig. 5.4. Control of System 5.83 (solid line - optimal PI control, dotted line - Smith and Corripio)

The PI parameters for the controller are

$$\left.\begin{array}{l} k_i(k) = 0.0036 \\ k_p(k) = 0.4045 - 0.0894 \times 0.9911^k \end{array}\right\} \; 0 \le k \le h, \qquad (5.87)$$

$$\left.\begin{array}{l} k_i(k) = 0.1735 \\ k_p(k) = 0.0015 \end{array}\right\} \; k \ge h, \qquad (5.88)$$

while the control parameters according to the method of Smith and Corripio (1985) are

$$k_i(k) = 0.3382, \qquad (5.89)$$

$$k_p(k) = 0.0034. \qquad (5.90)$$

Again, the simulation result in Fig. 5.5 shows a great improvement of the controller over the method presented by Smith and Corripio (1985).

It can be seen from these examples that the optimal PI controller can provide

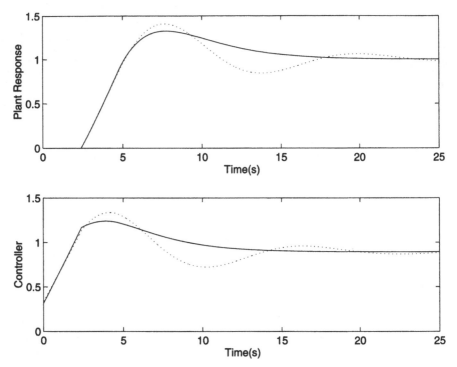

Fig. 5.5. Control of System 5.85 (solid line - optimal PI control, dotted line - Smith and Corripio)

good compensation for delay systems, even in the presence of small modelling perturbations, resulting in essentially invariant closed-loop responses.

5.2.11 Real-time Experiments

The proposed PI controller is applied in a real-time experiment to a *KI-100 Dual Process Simulator* from Kent Ridge Instruments. This is an analogue system simulator which can be configured to simulate a wide range of industrial systems with different kinds of dynamics and at different levels of noise. The simulator is connected to a PC via an A/D and D/A board, *LABVIEW* 4.1 from National Instruments. The sampling period used in the software is 0.01s.

Initially, from $t = 0$ to $t = 15$, a model identification experiment using relay feedback (Åström and Hägglund, 1995) is conducted, based on which, the first-order model for the system is identified as

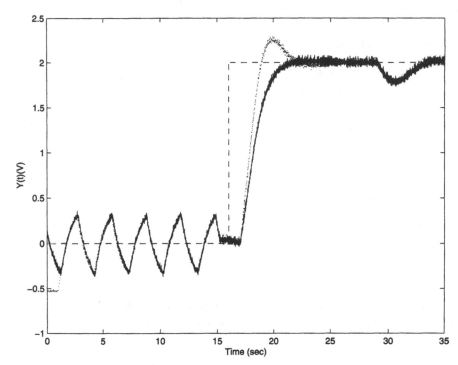

Fig. 5.6. Real-time experiment (solid line - optimal PI control, dotted line - Smith and Corripio)

$$G(s) = \frac{0.995}{s + 0.92} e^{-1.00s}. \tag{5.91}$$

A set-point change occurs at $t = 16$, and a step load disturbance occurs at $t = 29$. The performance of the control system is given in Fig. 5.6. For comparison, a PI controller tuned using Smith and Corripio's method (1985) is employed with the same system. The corresponding closed-loop response is shown on the same figure.

5.3 Optimal PID Controller Design Based on GPC

The PI controller designed based on GPC can be extended to full PID when a second-order model is considered.

5.3.1 General System Model

Consider a general second-order discrete-time transfer function:

$$G'(z) = \frac{b'_1 z + b'_2}{z^2 + a'_1 z + a'_2} z^{-h}. \tag{5.92}$$

Letting $e = y_{sp} - y$, the equivalent time-domain difference equation of (5.92) is given in (5.93) with $y_{sp} = 0$.

$$e(k+2) = -a'_1 e(k+1) - a'_2 e(k) - b'_1 u(k+1-h) - b'_2 u(k-h). \tag{5.93}$$

The model described by (5.93) encompasses a rich and general class of time-continuous linear systems given by

$$G(s) = \frac{ds + c}{(s+a)(s+b)} e^{-sL}. \tag{5.94}$$

In time-discrete form, the pulse-transfer function of (5.94) is adequately represented by the generic form of (5.95), where the parameters are related by the following equations:

$$
\begin{aligned}
a'_1 &= -(e^{-aT_s} + e^{-bT_s}), \\
a'_2 &= e^{-(a+b)T_s}, \\
b'_1 &= \frac{c}{ab}\left[\frac{ae^{-bT_s} - be^{-aT_s}}{b-a} + 1\right] + d\left[\frac{e^{-aT_s} - e^{-bT_s}}{b-a}\right], \\
b'_2 &= \frac{c(a-b)e^{-(a+b)T_s} + a(bd-c)e^{-aT_s} + b(c-ad)e^{-bT_s}}{ab(a-b)}, \\
h &= \frac{L}{T_s},
\end{aligned}
\tag{5.95}
$$

where T_s is the sampling interval.

From (5.93), shifting the time origin results in

$$e(k+1) = -a'_1 e(k) - a'_2 e(k-1) - b'_1 \tilde{u}(k-h), \tag{5.96}$$

where $\tilde{u}(k-h) = u(k-h) + \frac{b'_2}{b'_1} u(k-h-1)$.

In the following derivation, a GPC approach is used to derive the general control law. Then based on this GPC control law, the equivalent set of PID control parameters will be back calculated. Formulating (5.96) using a state-space description, gives:

$$
\begin{pmatrix} e(k) \\ e(k+1) \\ \theta(k+1) \end{pmatrix} = \begin{pmatrix} 0 & 1 & 0 \\ -a'_2 & -a'_1 & 0 \\ 0 & 1 & 1 \end{pmatrix} \begin{pmatrix} e(k-1) \\ e(k) \\ \theta(k) \end{pmatrix} + \begin{pmatrix} 0 \\ -b'_1 \\ 0 \end{pmatrix} \tilde{u}(k-h).
$$

Defining

$$F = \begin{pmatrix} 0 & 1 & 0 \\ -a_2' & -a_1' & 0 \\ 0 & 1 & 1 \end{pmatrix}, B = \begin{pmatrix} 0 \\ -b_1' \\ 0 \end{pmatrix}, X(k) = \begin{pmatrix} e(k) \\ e(k+1) \\ \theta(k+1) \end{pmatrix},$$

it follows that

$$X(k+1) = FX(k) + B\tilde{u}(k-h). \tag{5.97}$$

Using m control and p output prediction horizons, the future outputs $X(k+l)$ can be obtained recursively as follows (where $p \geq m$):

$$
\begin{aligned}
X(k+1) \quad &= FX(k) + B\tilde{u}(k-h), \\
X(k+2) \quad &= F^2 X(k) + FB\tilde{u}(k-h) + B\tilde{u}(k+1-h), \\
\cdots \quad &= \cdots \\
X(k+p-1) &= F^{p-1}X(k) + F^{p-2}B\tilde{u}(k-h) + \cdots \\
&\quad + F^{p-m-1}B\tilde{u}(k+m-2-h), \\
X(k+p-1) &= F^p X(k) + F^{p-1}B\tilde{u}(k-h) + \cdots \\
&\quad + F^{p-m}B\tilde{u}(k+m-1-h).
\end{aligned}
$$

Combining the equations, the following augmented matrix description is obtained:

$$\bar{X} = GFX(k) + A\tilde{U}, \tag{5.98}$$

where

$$\bar{X} = (X^T(k+1)\ X^T(k+2)\ \cdots\ X^T(k+p))^T,$$

$$G = \begin{pmatrix} I \\ F \\ \vdots \\ F^{p-1} \end{pmatrix},$$

$$A = \begin{pmatrix} B & 0 & \cdots & 0 \\ FB & B & \cdots & \vdots \\ \vdots & \vdots & \ddots & 0 \\ F^{p-1}B & F^{p-2}B & \cdots & F^{p-m}B \end{pmatrix},$$

$$\tilde{U} = (\tilde{u}(k-h)\ \tilde{u}(k+1-h)\ \cdots\ \tilde{u}(k+m-1-h))^T.$$

5.3.2 GPC Control Design

Consider the following quadratic cost function:

$$J = \sum_{l=1}^{p} \|x(k+l)\|^2_{Q(l)} + \sum_{j=1}^{m} \|u(k+j-1)\|^2_{R(j)}. \tag{5.99}$$

The main idea of a GPC design, as in the PI control case, is to derive a series of m controls at each sample time so that the cost function is minimised. Substituting (5.98) into (5.99), the cost function can be rewritten as follows:

$$J = [GFX(k) + A\tilde{U}]^T Q[GFX(k) + A\tilde{U}] + \tilde{U}^T R\tilde{U}. \tag{5.100}$$

In (5.100), Q is the state weighting matrix and R is the weighting vector for the control effort. The solution that minimises the cost function J can be obtained by solving (5.101):

$$\frac{\delta J}{\delta \tilde{U}} = 2A^T Q[GFX(k) + A\tilde{U}] + 2R\tilde{U} = 0. \tag{5.101}$$

Thus, the optimal control sequence is given as

$$\tilde{U} = -[A^T QA + R]^{-1}[A^T QGF]X(k). \tag{5.102}$$

Under the principle of receding horizon, only the first value of the optimal control sequence is output at each sampling time step. Thus, (5.102) can be rewritten as

$$\begin{aligned} \tilde{u}(k-h) &= -H[A^T QA + R]^{-1}[A^T QGF]X(k) \\ &= -DX(k) \end{aligned}, \tag{5.103}$$

where $D = -H[A^T QA + R]^{-1}[A^T QGF]$ and $H = [1\ 0\ \dots\ 0]$.

It follows that

$$\tilde{u}(k) = -DX(k+h).$$

In (5.103), the current control value is a linear combination of the future predicted states in $X(k+h)$. The future states needs to be predicted from the present ones, in order to have a causal control law. From (5.103) and (5.97), the following equations may be obtained:

$$X(k+1) = FX(k), \qquad 0 \leq k < h, \tag{5.104}$$
$$X(k+1) = (F - BD)X(k), \qquad k \geq h. \tag{5.105}$$

Lemma 5.1 extends the prediction to h steps.

Lemma 5.1

The future states of the system described by (5.97) are given by:

$$X(k+h) = (F - BD)^k F^{h-k} X(k), \; 0 \leq k < h, \tag{5.106}$$
$$X(k+h) = (F - BD)^h X(k), \; k \geq h. \tag{5.107}$$

Proof

For $k \geq h$, it is directly obtained from (5.105) recursively that

$$X(k+h) = (F - BD)^h X(k).$$

(5.107) is proven. (5.106) for $0 \leq k < h$ will be proven using an induction method.

For $k = 0$, (5.104) results in

$$X(h) = F^h X(0).$$

For $k = 1$, from (5.105),

$$X(1+h) = (F - BD)X(h).$$

Substituting $X(h) = F^{h-1}X(1)$ according to (5.104) gives

$$X(1+h) = (F - BD)F^{h-1}X(1).$$

(5.106) is proven for $k = 0, 1$.

Assume next that (5.106) holds for $k = j < h - 1$, i.e.,

$$X(j+h) = (F - BD)^j F^{h-j} X(j).$$

It will be proven that (5.106) holds also for $k = j + 1 < h$.

$$X(j + 1 + h) = (F - BD)X(j + h)$$
$$= (F - BD)(F - BD)^j F^{h-j} X(j).$$

Now, $X(j) = F^{-1}X(j+1)$ according to (5.104). Therefore,

$$X(j + 1 + h) = (F - BD)^{j+1} F^{h-(j+1)} X(j+1).$$

(5.106) is proven for $k = j + 1$. The lemma is proven by induction.

Denoting

$$[K_1(k)\ K_2(k)\ K_3(k)] = D(F - BD)^k F^{h-k},$$

the final control laws are given as follows:

$$\tilde{u} = -D(F - BD)^k F^{h-k} X(k)\quad for\ 0 \le k < h$$
$$= -[K_1(k)\ K_2(k)\ K_3(k)]X(k)$$
$$= -[(K_1(k) + K_2(k))e(k) - K_1(k)(e(k) - e(k-1)) + K_3(k)\theta(k)],$$
$$\tilde{u} = -D(F - BD)^h X(k)\quad for\ k \ge h$$
$$= -[(K_1(h) + K_2(h))e(k) - K_1(h)(e(k) - e(k-1)) + K_3(h)\theta(k)].$$

This results in the following theorem.

Theorem 5.11

For the system described by (5.93), the GPC control is given by

$$\tilde{u}(k) = K_P(k)e(k) + K_I(k)\theta(k) + K_D(k)(e(k) - e(k-1)), \qquad (5.108)$$

where

$$\left.\begin{aligned} K_P &= -(K_1(k) + K_2(k)) \\ K_I &= -K_3(k) \\ K_D &= K_1(k) \end{aligned}\right\} \quad 0 \le k < h,$$

and

$$\left.\begin{aligned} K_P &= -(K_1(h) + K_2(h)) \\ K_I &= -K_3(h) \\ K_D &= K_1(h) \end{aligned}\right\} \quad k \ge h.$$

Anti-integral-windup measures may be incorporated directly into the control laws to improve performance of time-delay systems. For these systems, there is no response during the time-delay dead zone ($0 \leq k < h$). It is desirable to avoid error accumulation during this period of time, otherwise the accumulation will lead to backlash in terms of a large control effort, causing a large response overshoot and long settling time.

To this end, let

$$X(k+1) = X(k), \qquad 0 \leq k < h.$$

It follows that

$$X(k+h) = (F - BD)^k X(h) = (F - BD)^k X(0).$$

This results in the following corollary.

Corollary 5.1

For the system described by (5.93), the GPC control with anti-windup measures is given by

$$\begin{aligned}
\tilde{u}(k) &= K_P(k)e(k) + K_I(k)\theta(k) + K_D(k)(e(k) - e(k-1)) \\
& \qquad 0 \leq k < h, \quad\quad\quad\quad\quad\quad\quad\quad\quad\quad\quad\quad\quad\quad (5.109) \\
\tilde{u}(k) &= K_P(k)e(k) + K_I(k)\theta(k) + K_D(k)(e(k) - e(k-1)) \\
& \qquad k \geq h, \quad\quad\quad\quad\quad\quad\quad\quad\quad\quad\quad\quad\quad\quad\quad\quad (5.110)
\end{aligned}$$

where $K_P(k)$, $K_I(k)$ and $K_D(k)$ are defined as in Theorem 5.11.

5.3.3 Selection of Control Horizon

The control horizon, m, is chosen such that it equals the prediction horizon, p. In addition, as in the PI control case, let $\tilde{u}(k+1) = \tilde{u}(k+2) = \ldots = \tilde{u}(k+m)$, this allows for $(m-1)$ steps of infinite weighting on the changes of the control action ahead of one step of finite weight. This choice is valid as long as the system is stable with fairly damped poles as shown by Clarke *et al.* (1987). It is naturally robust against unmodelled dynamics and attractive for control applications. The matrices A may be replaced by the following relation:

$$\begin{aligned}
A\tilde{U} &= \begin{pmatrix} B \\ FB + B \\ \vdots \\ F^{p-1}B + \cdots + B \end{pmatrix} \tilde{u}(k-h) \\
&= A_o B \tilde{u}(k-h).
\end{aligned} \qquad (5.111)$$

The control laws are thus given by

$$\tilde{u}(k) = \begin{cases} -D(F - BD)^k F^{h-k} X(k), & 0 \le k < h, \\ -D(F - BD)^h X(k), & k \ge h, \end{cases}$$

where $D = -H[(A_o B)^T Q(A_o B) + R]^{-1}[(A_o B)^T Q G F] X(k)$.

5.3.4 Selection of Weighting Matrices

As in the PI control case, it is useful to link the selection of weighting matrices to desirable classical specifications, i.e., a direct relationship between the weighting matrices Q and R, and classical specifications such as the damping ratio ζ and natural frequency ω_n of the closed-loop system. These relationships are provided in this section.

Consider (5.97) and (5.103). When $k \ge h$, the closed-loop equation is given as follows:

$$X(k+1) = (F - BD)X(k). \tag{5.112}$$

Let

$$D = [d_1 \ d_2 \ d_3],$$

the closed-loop characteristic equation is thus given by

$$
\begin{aligned}
det[zI - (F - BD)] &= det \begin{pmatrix} z & -1 & 0 \\ a_2' - b_1' d_1 & z - (b_1' d_1 - a_1') & -b_1' \\ 0 & -1 & z-1 \end{pmatrix} \\
&= z^3 + (a_1' - 1 - b_1' d_2)z^2 \\
&\quad + [a_2' - a_1' - b_1'(d_1 - d_2 + d_3)]z + b_1' d_1 - a_2'.
\end{aligned}
$$

D can be chosen to achieve a desired closed-loop pole constellation. Under a dominant pole placement design, the third pole can be placed more than four times deeper in the left-hand plane, so that the response is dominated by the two complex poles. The desired characteristic equation in the s-domain is thus given by

$$(s^2 + 2\zeta \omega_n s + \omega_n^2)(s + p_1) = 0, \tag{5.113}$$

where ζ and ω_n are the damping ratio and natural frequency specifications of the closed loop, and p_1 corresponds to the third pole. This is equivalent to

$$z^3 + (a_1 - e^{-p_1 T_s})z^2 + (a_2 - a_1 e^{-p_1 T_s})z - a_2 e^{-p_1 T_s} = 0 \qquad (5.114)$$

in the z-domain, where

$$\omega = \omega_n \sqrt{1 - \zeta^2},$$
$$\alpha = e^{-\zeta \omega_n T_s}, \beta = \cos(\omega T_s), \gamma = \sin(\omega T_s),$$
$$a_1 = -2\alpha\beta, a_2 = \alpha^2.$$

Equating the coefficients of (5.113) and (5.114), the elements of D may be obtained as follows:

$$\begin{aligned}
d_1 &= \frac{a_2' - a_2 e^{-p_1 T_s}}{b_1'}, \\
d_2 &= \frac{a_1' - a_1 - 1 + e^{-p_1 T_s}}{b_1'}, \\
d_3 &= \frac{(1 + a_1 + a_2)(e^{-p_1 T_s} - 1)}{b_1'}.
\end{aligned} \qquad (5.115)$$

Thus, for time $k \geq h$, the D elements ($d_1 \ d_2 \ d_3$) required to achieve closed-loop specifications ζ, ω_n and p_1 can be obtained from (5.115).

5.3.5 Choice of ζ, ω_n and P_1

The specification ζ relates to the damping ratio of the closed loop. For a well-damped response, generally $\zeta > 0.7$ is desired. With ζ fixed, the choice of natural frequency ω_n will determine the closed-loop response speed. The following approximate relationship between ζ, ω_n and the desired closed-loop time constant τ_{cl} provides a good guideline for the choice of ω_n.

The choice of ζ and ω_n is essentially to achieve a compromise between the often conflicting control objectives of speed and stability. Finally, the third pole may be placed at least four times deeper into the left-hand plane, so there is no dominant effect on the closed-loop dynamics. The pole p_1 can thus be specified automatically according to

$$p_1 = r \frac{1}{\zeta \omega_n}, r > 4.$$

5.3.6 Q–D Relationship

With D thus computed according to the classical specifications, it will be interesting to determine the equivalent GPC weights in Q. This relationship can yield an initial weight set from which finer adjustment of the GPC parameters can be made for enhanced performance. The relationship between D, Q and R is given in Theorem 5.12.

Theorem 5.12

The equivalent GPC Q matrix is related to d_1, d_2, d_3 and R by

$$
\begin{pmatrix} q_1 \\ q_2 \\ q_3 \end{pmatrix} = \begin{pmatrix} \kappa_1 & \kappa_2 & \kappa_3 \\ \lambda_1 & \lambda_2 & \lambda_3 \\ \mu_1 & \mu_2 & \mu_3 \end{pmatrix}^{-1} \begin{pmatrix} d_1 \\ -d_2 \\ -d_3 \end{pmatrix} R,
\tag{5.116}
$$

where the notation is defined in the following proof.

Proof

In (5.103), D is given in terms of matrices G, F, A, R and Q. The state weighting matrix Q has a form given in (5.117). The size of Q depends on the choice of prediction horizon p.

$$
Q = \begin{pmatrix}
q_1 & 0 & 0 & \cdots & 0 & 0 & 0 \\
0 & q_2 & 0 & \cdots & & & \\
0 & 0 & q_3 & \cdots & \vdots & \vdots & \vdots \\
& & & \ddots & & & \\
\vdots & \vdots & \vdots & & q_1 & 0 & 0 \\
0 & & \vdots & & 0 & q_2 & 0 \\
0 & 0 & 0 & \cdots & 0 & 0 & q_3
\end{pmatrix}_{3_p \times 3_p}
\tag{5.117}
$$

$$
= \begin{pmatrix}
Q_o & 0 & \cdots & 0 \\
0 & Q_o & \cdots & 0 \\
\vdots & \vdots & \ddots & \vdots \\
0 & \cdots & \cdots & Q_o
\end{pmatrix}, \text{ where } \quad Q_o = \begin{pmatrix} q_1 & 0 & 0 \\ 0 & q_2 & 0 \\ 0 & 0 & q_3 \end{pmatrix}.
$$

G, F and A are constant matrices in terms of system parameters only. Consider D. The first part is a scalar and the second part a 1×3 vector

$$
\begin{aligned}
D &= -[(A_o B)^T Q (A_o B) + R]^{-1} [(A_o B)^T Q G F] \\
&= -[B^T A_o^T Q A_o B + R]^{-1} [B^T A_o^T Q G F].
\end{aligned}
\tag{5.118}
$$

Expanding the common factor $A_o^T Q A_o$, it follows that

$$
A_o^T Q A_o = [I \ (F+I)^T \ \cdots \ (F^{p-1} + \cdots + I)^T]
$$

$$\times \begin{bmatrix} Q_o & 0 & \cdots & 0 \\ 0 & Q_o & & \vdots \\ \vdots & & \ddots & \vdots \\ 0 & \cdots & \cdots & Q_o \end{bmatrix} \begin{bmatrix} I \\ F+I \\ \vdots \\ F^{p-1} + \cdots + I \end{bmatrix}$$

$$= Q_o + (F+I)^T Q_o (F+I) + (F^2 + F + I)^T Q_o (F^2 + F + I)$$
$$+ \cdots + (F^{p-1} + \cdots + I)^T Q_o (F^{p-1} + \cdots + I)$$
$$= Q_{01} + (F+I)^T Q_{01} (F+I)$$
$$+ (F^2 + F + I)^T Q_{01} (F^2 + F + I)$$
$$+ \cdots + (F^{p-1} + \cdots + I)^T Q_{01} (F^{p-1} + \cdots + I)$$
$$+ Q_{02} + (F+I)^T Q_{02} (F+I)$$
$$+ (F^2 + F + I)^T Q_{02} (F^2 + F + I)$$
$$+ \cdots + (F^{p-1} + \cdots + I)^T Q_{02} (F^{p-1} + \cdots + I)$$
$$+ Q_{03} + (F+I)^T Q_{03} (F+I)$$
$$+ (F^2 + F + I)^T Q_{03} (F^2 + F + I)$$
$$+ \cdots + (F^{p-1} + \cdots + I)^T Q_{03} (F^{p-1} + \cdots + I),$$

where

$$Q_{01} = \begin{pmatrix} q_1 & 0 & 0 \\ 0 & 0 & 0 \\ 0 & 0 & 0 \end{pmatrix}, Q_{02} = \begin{pmatrix} 0 & 0 & 0 \\ 0 & q_2 & 0 \\ 0 & 0 & 0 \end{pmatrix}, Q_{03} = \begin{pmatrix} 0 & 0 & 0 \\ 0 & 0 & 0 \\ 0 & 0 & q_3 \end{pmatrix}.$$

Consider a general term,

$$(F^T + \cdots + I)^n Q_{0k} (F^T + \cdots + I) = q_k$$
$$\times \begin{pmatrix} \times & \times & \times \\ f_{k2}^{(n)} f_{k1}^{(n)} & f_{k2}^{(n)} f_{k2}^{(n)} & f_{k2}^{(n)} f_{k3}^{(n)} \\ \times & \times & \times \end{pmatrix}$$
$$\text{for } k = 1, 2, 3. \tag{5.119}$$

In (5.119), $f_{ij}^{(n)}$ is the ith row and jth column element of $F^n + \cdots + I$. Therefore, it follows that

$$\sum_{n=1}^{p-1} (F^T + \cdots + I)^n Q_{0k} (F^T + \cdots + I) = q_k$$
$$\times \begin{pmatrix} \times & \times & \times \\ \sum_{n=1}^{p-1} f_{k2}^{(n)} f_{k1}^{(n)} & \sum_{n=1}^{p-1} f_{k2}^{(n)} f_{k2}^{(n)} & \sum_{n=1}^{p-1} f_{k2}^{(n)} f_{k3}^{(n)} \\ \times & \times & \times \end{pmatrix}. \tag{5.120}$$

Thus, $A_o^T Q A_o$ may be expressed as:

$$A_o^T Q A_o = \begin{pmatrix} \times & \times & \times \\ \alpha & \beta & \gamma \\ \times & \times & \times \end{pmatrix}, \tag{5.121}$$

where

$$
\begin{aligned}
\alpha &= \sum_{k=1}^{3} q_k \left(\sum_{n=1}^{p-1} f_{k2}^{(n)} f_{k1}^{(n)} \right) \\
&= q_1 \sum_{n=1}^{p-1} f_{12}^{(n)} f_{11}^{(n)} + q_2 \sum_{n=1}^{p-1} f_{22}^{(n)} f_{21}^{(n)} + q_3 \sum_{n=1}^{p-1} f_{32}^{(n)} f_{31}^{(n)} \\
&= \alpha_1 q_1 + \alpha_2 q_2 + \alpha_3 q_3, \\
\beta &= \sum_{k=1}^{3} q_k \left(\sum_{n=1}^{p-1} f_{k2}^{(n)} f_{k2}^{(n)} \right) + q_2 \\
&= q_1 \sum_{n=1}^{p-1} f_{12}^{(n)} f_{12}^{(n)} + q_2 (1 + \sum_{n=1}^{p-1} f_{22}^{(n)} f_{22}^{(n)}) + q_3 \sum_{n=1}^{p-1} f_{32}^{(n)} f_{32}^{(n)} \\
&= \beta_1 q_1 + \beta_2 q_2 + \beta_3 q_3, \\
\gamma &= \sum_{k=1}^{3} q_k \left(\sum_{n=1}^{p-1} f_{k2}^{(n)} f_{k3}^{(n)} \right) \\
&= q_1 \sum_{n=1}^{p-1} f_{12}^{(n)} f_{13}^{(n)} + q_2 \sum_{n=1}^{p-1} f_{22}^{(n)} f_{23}^{(n)} + q_3 \sum_{n=1}^{p-1} f_{32}^{(n)} f_{33}^{(n)} \\
&= \gamma_1 q_1 + \gamma_2 q_2 + \gamma_3 q_3.
\end{aligned}
$$

Similarly, $A_o^T Q G$ can be expressed as:

$$A_o^T Q G = \begin{pmatrix} \times & \times & \times \\ \bar{\alpha} & \bar{\beta} & \bar{\gamma} \\ \times & \times & \times \end{pmatrix},$$

where $g_{ij}^{(n)}$ is the ith row and jth column element of F^n and

$$
\begin{aligned}
\bar{\alpha} &= \sum_{k=1}^{3} q_k \left(\sum_{n=1}^{p-1} f_{k2}^{(n)} g_{k1}^{(n)} \right) \\
&= q_1 \sum_{n=1}^{p-1} f_{12}^{(n)} g_{11}^{(n)} + q_2 \sum_{n=1}^{p-1} f_{22}^{(n)} g_{21}^{(n)} + q_3 \sum_{n=1}^{p-1} f_{32}^{(n)} g_{31}^{(n)} \\
&= \bar{\alpha}_1 q_1 + \bar{\alpha}_2 q_2 + \bar{\alpha}_3 q_3, \\
\bar{\beta} &= \sum_{k=1}^{3} q_k \left(\sum_{n=1}^{p-1} f_{k2}^{(n)} g_{k2}^{(n)} \right) + q_2 \\
&= q_1 \sum_{n=1}^{p-1} f_{12}^{(n)} g_{12}^{(n)} + q_2 (1 + \sum_{n=1}^{p-1} f_{22}^{(n)} g_{22}^{(n)}) + q_3 \sum_{n=1}^{p-1} f_{32}^{(n)} g_{32}^{(n)} \\
&= \bar{\beta}_1 q_1 + \bar{\beta}_2 q_2 + \bar{\beta}_3 q_3, \\
\bar{\gamma} &= \sum_{k=1}^{3} q_k \left(\sum_{n=1}^{p-1} f_{k2}^{(n)} f_{k3}^{(n)} \right) \\
&= q_1 \sum_{n=1}^{p-1} f_{12}^{(n)} g_{13}^{(n)} + q_2 \sum_{n=1}^{p-1} f_{22}^{(n)} g_{23}^{(n)} + q_3 \sum_{n=1}^{p-1} f_{32}^{(n)} g_{33}^{(n)} \\
&= \bar{\gamma}_1 q_1 + \bar{\gamma}_2 q_2 + \bar{\gamma}_3 q_3.
\end{aligned}
$$

Hence, the two components in D may be expressed as

$$B^T A_o^T QGF = [0 \ -b_1' \ 0] \begin{bmatrix} \times & \times & \times \\ \bar{\alpha} & \bar{\beta} & \bar{\gamma} \\ \times & \times & \times \end{bmatrix} \begin{bmatrix} 0 & 1 & 0 \\ -a_2' & -a_2' & 0 \\ 0 & 1 & 1 \end{bmatrix}$$

$$= -b_1' \, (-a_2' \bar{\alpha} \, \bar{\alpha} - a_1' \bar{\beta} + \bar{\gamma} \, \bar{\gamma}),$$

$$B^T A^T QAB = [0 \ -b_1' \ 0] \begin{bmatrix} \times & \times & \times \\ \alpha & \beta & \gamma \\ \times & \times & \times \end{bmatrix} \begin{bmatrix} 0 \\ -b_1' \\ 0 \end{bmatrix}$$

$$= b_1'^2 \beta.$$

(5.122)

Combining the two parts of (5.122), it follows that

$$D = -\frac{b_1' \, [-a_2' \bar{\alpha} \, \bar{\alpha} - a_1' \bar{\beta} + \bar{\gamma} \, \bar{\gamma}]}{b_1'^2 + R} = [d_1 \ d_2 \ d_3].$$

(5.123)

Expressing (5.123) in terms of q_1, q_2 and q_3,

$$(b_1' a_2' \bar{\beta}_1 - d_1 b_1'^2 \beta_1)q_1 + (b_1' a_2' \bar{\beta}_2 - d_1 b_1'^2 \beta_2)q_2 + (b_1' a_2' \bar{\beta}_3 - d_1 b_1'^2 \beta_3)q_3$$
$$= d_1 R,$$
$$[b_1'(\bar{\alpha}_1 + \bar{\gamma}_1 + d_2 b_1' \beta_1 - a_1' \bar{\beta}_1)]q_1 + [b_1'(\bar{\alpha}_2 + \bar{\gamma}_2 + d_2 b_1' \beta_2 - a_1' \bar{\beta}_2)]q_2$$
$$+ [b_1'(\bar{\alpha}_3 + \bar{\gamma}_3 + d_2 b_1' \beta_3 - a_1' \bar{\beta}_3)]q_3 = -d_2 R,$$
$$(b_1' \bar{\gamma}_1 + d_3 b_1'^2 \beta_1)q_1 + (b_1' \bar{\gamma}_2 + d_3 b_1'^2 \beta_2)q_2 + (b_1' \bar{\gamma}_3 + d_3 b_1'^2 \beta_3)q_3 = -d_3 R.$$

Using a matrix formulation,

$$\begin{pmatrix} \kappa_1 & \kappa_2 & \kappa_3 \\ \lambda_1 & \lambda_2 & \lambda_3 \\ \mu_1 & \mu_2 & \mu_3 \end{pmatrix} \begin{pmatrix} q_1 \\ q_2 \\ q_3 \end{pmatrix} = \begin{pmatrix} d_1 R \\ -d_2 R \\ -d_3 R \end{pmatrix},$$

where

$$\kappa_i = b_1' a_2' \bar{\beta}_1 - d_1 b_1'^2 \beta_1,$$
$$\lambda_i = b_1'(\bar{\alpha}_i + \bar{\gamma}_i + d_2 b_1' \beta_i - a_1' \bar{\beta}_i),$$
$$\mu_i = b_1' \bar{\gamma}_i + d_3 b_1'^2 \beta_i,$$

for $i = 1, 2, 3$. Thus, the weighting factors are given by:

$$\begin{pmatrix} q_1 \\ q_2 \\ q_3 \end{pmatrix} = \begin{pmatrix} \kappa_1 & \kappa_2 & \kappa_3 \\ \lambda_1 & \lambda_2 & \lambda_3 \\ \mu_1 & \mu_2 & \mu_3 \end{pmatrix}^{-1} \begin{pmatrix} d_1 \\ -d_2 \\ -d_3 \end{pmatrix} R.$$

(5.124)

The proof is completed.

Remark 5.5

There are four unknown parameter q_1, q_2, q_3 and R, but only three equations in (5.116). However, it is only the relative magnitudes of these parameters which matter to the cost function (5.99). The absolute value of R is not directly useful and significant. In practice, it may be set to unity.

Remark 5.6

The user may thus choose to provide classical control specifications in terms of ζ, ω_n and p_1, and the equivalent weighting matrices can be computed according to (5.115) and (5.116). The equivalent weights may be used directly to commission the PID control, or they may serve as initial values for subsequently direct fine tuning of the weights by the operator.

5.3.7 Simulation

In this section several simulation examples, involving systems of different dynamics, are given to demonstrate the wide applicability of the control algorithm. The common specification $p_1 = 20$ is assumed. In addition, the sampling time used was 1ms. Additive white noise is also included to simulate measurement noise.

Example 5.4

Consider the second-order system:

$$G(s) = \frac{\omega_p^2}{(s^2 + 2\zeta_p\omega_p s + \omega_p^2)}, \tag{5.125}$$

with $\omega_p = 10$ and $\zeta_p = 0.5$ and 5 to provide under damped and over damped characteristics for simulation. The specifications $\zeta = 2$, $\omega_n = 10$ are assumed for both. The responses are shown in Fig. 5.7 and Fig. 5.8. Note both closed-loop response are rather similar, although the open-loop damping ratio differs by a factor of 10. The PID gains for the underdamped and overdamped system are 7.788, 0.01951, 485.3 and 8.819, 0.0204, −388 respectively.

Example 5.5

Consider the unstable system given by

$$G(s) = \frac{-1}{(s - 3)(s + 1)}. \tag{5.126}$$

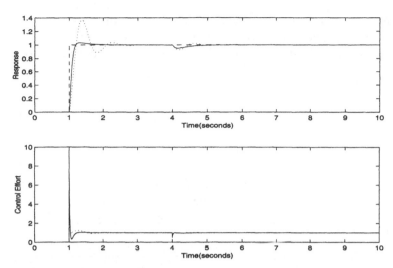

Fig. 5.7. Simulated response for an underdamped system $\zeta_p = 0.5$, $\zeta = 2$, $\omega_n = 10$ with the dotted line being the counterpart of the ZN tuned controller

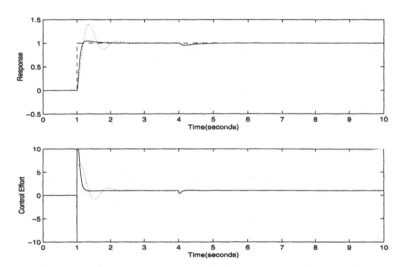

Fig. 5.8. Simulated response for an underdamped system $\zeta_p = 5$, $\zeta = 2$, $\omega_n = 10$ with the dotted line being the counterpart of the ZN tuned controller

The closed-loop response is shown in Fig. 5.9 with the specifications $\zeta = 2$ and $\omega_n = 2.5$. The PID gains were 206, 0.123 and 31520. An input step disturbance was added at $t = 10$s. Next consider an unstable system with

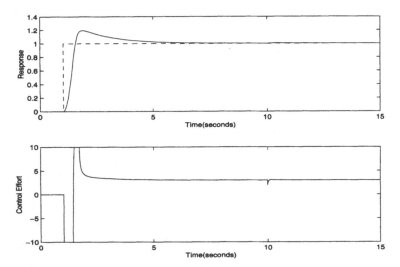

Fig. 5.9. Simulated response of unstable system $\zeta = 2$, $\omega_n = 2.5$

dead time:

$$G(s) = \frac{1}{(s - 0.1)}e^{-0.2s}. \tag{5.127}$$

The closed-loop response is shown in Fig. 5.10 with specification $\zeta = 2$, $\omega_n = 3$. The time-varying gains are shown in Fig. 5.11. A 10% input step disturbance was applied at $t = 4$s.

Example 5.6

Thus far, simulations have been done for systems with a structure which is contained completely in the system model (5.93). It is interesting to see how control system with the same model (5.93) performs when faced with systems of dissimilar structures. This simulation will also show the robustness of the controller in the face of non-parametric modelling uncertainties.

Consider the high-order system:

$$G(s) = \frac{1}{(s + 1)^{10}}. \tag{5.128}$$

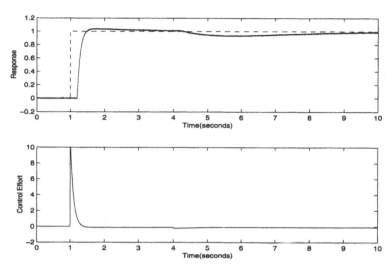

Fig. 5.10. Simulated response of unstable system $\zeta = 2$, $\omega_n = 3$

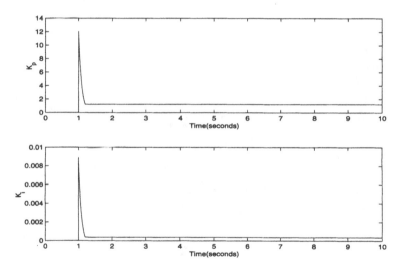

Fig. 5.11. Variation of PID gains

A system identification experiment is carried out to obtain the parameters of (5.93). The response and control effort is given in Fig. 5.12 and the variation of controller gains is depicted in Fig. 5.13. A 10% input step disturbance is added at time 30s in the simulation using the present control method. A comparison with the method of Smith and Corripio (1985) is provided, with no input disturbance injected since the output has not settled down.

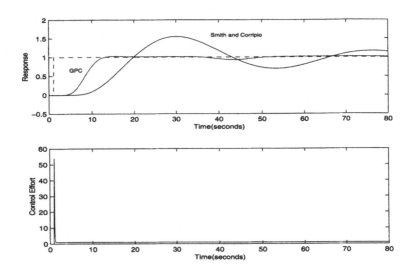

Fig. 5.12. Response and control effort for $n = 10$ ($\zeta = 5$, $\omega_n = 2$)

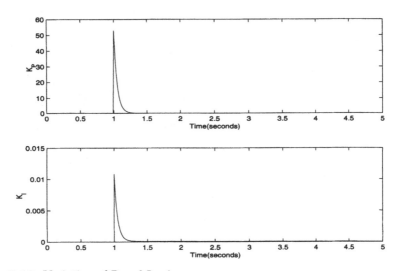

Fig. 5.13. Variation of P and I gains

CHAPTER 6
PREDICTIVE ITERATIVE LEARNING CONTROL

There are applications in which the control system is commanded to execute identical and repetitive tasks. These include pick-and-place robot control, moulding control, and many semiconductor processes. The iterative learning control (ILC) method first introduced by Arimoto *et al.* (1984) is based on the previous control history and a learning mechanism. The monograph by Moore (1992) contains more detail on the background. A recent book (Chen and Wen, 1999) surveys the literature on ILC up to 1998.

A predictive control mechanism can be incorporated into the ILC scheme to achieve optimisation over a horizon of future iterations of the same task, thus achieving performance which improved with time. This chapter is dedicated to the development of predictive iterative learning algorithms for linear and non-linear repetitive systems. A predictive iterative learning control (PILC) algorithm for linear systems is first studied, based on a package information model that represents the transition of the tracking error between two trials. The predictive algorithm is derived along the trial number (or repetition) index. The convergence and robustness of this algorithm is illustrated using a rigorous proof. However, this algorithm is based on package information, which occupies a lot of memory. Based on this consideration, a predictive iterative learning algorithm is discussed which does not need the use of package information. Examples are given to demonstrate the effectiveness of the two algorithms. For non-linear systems, a PILC algorithm is developed based on a neural network model.

In the framework to be developed below, the following norms are introduced:

$$\| f \| = \max_{1 \leq i \leq n} |f_i|, \quad \| S \| = \max_{1 \leq i \leq m} \left(\sum_{j=1}^{n} |s_{ij}| \right),$$

where $f = [f_1, ..., f_n]^T$ is a vector, and $S = [s_{ij}] \in R^{m \times n}$ is a matrix.

6.1 Linear Iterative Learning Control Algorithm

The concept of ILC for generating the optimal input to a system was first introduced by Arimoto and co-workers (1984). Figure 6.1 illustrates the basic idea. For each cycle, the system operates on the input and output signals ($u_k(t)$ and $y_k(t)$, respectively) which are stored in memory (some type of memory device is implicitly assumed in the block of Fig. 6.1 labelled "Learning Controller"). The learning control algorithm will then evaluate the performance error

$$e_k(t) = y_d(t) - y_k(t), \tag{6.1}$$

where $y_d(t)$ is the desired output of the system. Based on the error signal, the learning controller then computes a new input signal $u_{k+1}(t)$, which is stored for use during the next trial (by "the next trial", it means the next cycle of operation). The control action at the $(k+1)$th cycle may be defined as

$$u_{k+1}(t) = \mathcal{P}(u_k(t), e_k(t), t), \tag{6.2}$$

where the index k is a trial number. $U_{[T_0,T_f]}$ and $E_{[T_0,T_f]}$ are the input and error function space defined on $[T_0, T_f]$, respectively, and $\mathcal{P}$ is the operator such that $\mathcal{P} : U_{[T_0,T_f]} \times E_{[T_0,T_f]} \to U_{[T_0,T_f]}$. The next input command is chosen in such a way as to guarantee that the performance error will be reduced in the next trial.

Most of the popular designs for ILC can be roughly classified into D-type (Arimoto et al., 1984; Bien and Huh, 1989) and P-type ILC (Saab, 1994). Besides these, other systematic approaches such as optimal ILC (Amann et al., 1996) and predictive ILC along the time axis (Lee and Lee, 1999; Tan et al., 2002) have also been developed for linear and non-linear systems.

6.2 Predictive Iterative Learning Control Using Package Information

In this section, the incorporation of predictive features into the ILC using "package" information will be illustrated. The meaning of "package" information will be explained in due time.

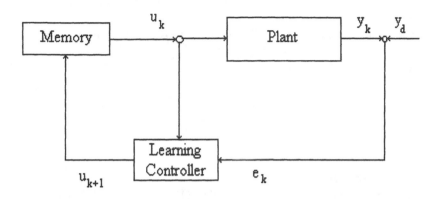

Fig. 6.1. Iterative learning control scheme

6.2.1 System Model and Problem Statements

Consider a repetitive linear discrete-time system with uncertainty and disturbance as follows:

$$x_i(t+1) = A(t)x_i(t) + B(t)u_i(t) + w_i(t), \qquad (6.3)$$
$$y_i(t) = H(t)x_i(t), \qquad (6.4)$$

where i denotes the ith repetitive operation of the system; $x_i(t) \in R^n, u_i(t) \in R^m$, and $y_i(t) \in R^r$ are the state, control input, and output of the system, respectively; $w_i(t)$ denotes the uncertainty or disturbance; $t \in [0, N]$ denotes the time; and $A(t), B(t)$, and $H(t)$ are matrices with appropriate dimensions.

The problem is stated as follows: Find an update mechanism for the input trajectory of a new repetition based on the information from the previous repetition operation so that the controlled output converges to the desired reference over time horizon $[0, N]$.

Due to the cyclic nature of the repetitive systems, it is convenient to pack the information in each cycle (or trial) together. The equation (6.3) becomes

$$x_i(1) = A(0)x_i(0) + B(0)u_i(0) + w_i(0),$$
$$x_i(2) = A(1)A(0)x_i(0) + A(1)B(0)u_i(0) + B(1)u_i(1)$$
$$+ A(1)w_i(0) + w_i(1),$$

$$\vdots$$

$$x_i(N) = \Pi_{k=0}^{N-1} A(k)x_i(0) + \Pi_{k=1}^{N-1} A(k)B(0)u_i(0) + \dots$$

$$+B(N-1)u_i(N-1) + \Pi_{k=1}^{N-1}A(k)w_i(0) + ... + w_i(N).$$

Pack the results in matrix form for all time steps of a repetition to obtain

$$\mathbf{y}_i = G\mathbf{u}_i + G_0 x_i(0) + \mathbf{w}_i, \tag{6.5}$$

where $\mathbf{y}_i = [y_i^T(1), y_i^T(2), ..., y_i^T(N)]^T$,

$G =$

$$\begin{bmatrix} H(1)B(0) & 0 & ... & 0 \\ H(2)A(1)B(0) & H(2)B(1) & ... & 0 \\ & \vdots & & \\ & & ... & \\ H(N)\Pi_{k=1}^{N-1}A(k)B(0) & H(N)\Pi_{k=2}^{N-1}A(k)B(1) & ... & H(N)B(N-1) \end{bmatrix},$$

$$G_0 = \begin{bmatrix} H(1)A(0) \\ H(2)A(1)A(0) \\ \vdots \\ H(N)\Pi_{k=0}^{N-1}A(k) \end{bmatrix},$$

$$\mathbf{w}_i = \begin{bmatrix} H(1)w_i(0) \\ H(2)A(1)w_i(0) + w_i(1) \\ \vdots \\ H(N)\Pi_{k=1}^{N-1}A(k)w_i(0) + ... + H(N)w_i(0) \end{bmatrix},$$

$$\mathbf{u}_i = [u_i^T(0), u_i^T(1), ..., u_i^T(N-1)]^T.$$

with $\mathbf{w}_i$ representing the batch-wise independent error (including measurement noises). G is an impulse response matrix which can be obtained through identification or linearisation of a non-linear model.

6.2.2 Predictor Construction

Once the state space model is available, the subsequent steps for predictor construction are straightforward. For this, the output error at the $(i+1)$th iteration can be written as

$$\begin{aligned} \mathbf{e}_{i+1} &= \mathbf{y}_d - \mathbf{y}_{i+1} \\ &= \mathbf{e}_i - (\mathbf{y}_{i+1} - \mathbf{y}_i) \\ &= \mathbf{e}_i - G\Delta\mathbf{u}_{i+1} - G_0\Delta x_{i+1}(0) - \Delta\mathbf{w}_{i+1}. \end{aligned} \tag{6.6}$$

The structure of error model (6.6) is useful in formulating predictive controllers. In order to do this, a prediction model is determined which is

$$\hat{\mathbf{e}}_{i+1} = \hat{\mathbf{e}}_i - G\Delta\mathbf{u}_{i+1}, \tag{6.7}$$

where $\hat{\mathbf{e}}_{i+1}$ denotes the error predicted at instant i for instant $i+1$. This model is redefined at each sampling instant i from the actual error vector previously applied, that is $\hat{\mathbf{e}}_i = \mathbf{e}_i$. Comparing (6.7) with (6.6), one may observe that (6.7) does not include the disturbance noise vector since it is assumed to be unknown. Also, $\Delta x_{i+1}(0)$ is not included in (6.7) as it will complicate the control formulation. By applying the equation (6.7) recursively, one may write:

$$\hat{\mathbf{e}}_{i+2} = \mathbf{e}_i - G\Delta\mathbf{u}_{i+2} - G\Delta\mathbf{u}_{i+1}, \tag{6.8}$$
$$\hat{\mathbf{e}}_{i+3} = \mathbf{e}_i - G(\Delta\mathbf{u}_{i+3} + \Delta\mathbf{u}_{i+2} + \Delta\mathbf{u}_{i+1}), \tag{6.9}$$
$$\vdots$$
$$\hat{\mathbf{e}}_{i+p} = \mathbf{e}_i - G(\Delta\mathbf{u}_{i+p} + \Delta\mathbf{u}_{i+p-1} + ... + \Delta\mathbf{u}_{i+1}), \tag{6.10}$$

where p is the prediction horizon.

6.2.3 Derivation of Algorithm

Consider a linear quadratic performance index written in the following form:

$$J = \sum_{j=1}^{p}[\hat{\mathbf{e}}_{i+j}^T\hat{\mathbf{e}}_{i+j} + \bar{\gamma}\Delta\mathbf{u}_{i+j}^T\Delta\mathbf{u}_{i+j}], \tag{6.11}$$

where $\bar{\gamma} > 0$ is the control weight and p is the prediction horizon. This criterion includes the error not only at the next trial, but also at the next p trials, as well as the corresponding changes in input. The weight parameter $\bar{\gamma} > 0$ determines the importance of more distant (future) errors and incremental inputs compared with the present ones. By including more future signals, the learning algorithm becomes less 'short sighted'(Amann et al., 1998).

The complexity introduced by the equations (6.7)–(6.10) is basically the result of the number of unknown control sequence Δu_{i+j}. One way of reducing the number of unknowns is to predetermine the form of the control sequence. It has proven useful to impose a step control sequence together with a cost function such as the one given in Chapter 4, thus reducing the number of unknowns to a single one (Clarke et al., 1987; Huang et al, 1999b). The control sequence is made to be constant over the prediction interval, i.e.,

$\Delta\mathbf{u}_{i+1} = \Delta\mathbf{u}_{i+2} = ... = \Delta\mathbf{u}_{i+p}$. Now the prediction equation with a $i+p$ horizon is

$$\hat{\mathbf{e}}_{i+1} = \mathbf{e}_i - G\Delta\mathbf{u}_{i+1}, \qquad (6.12)$$
$$\hat{\mathbf{e}}_{i+2} = \mathbf{e}_i - 2G\Delta\mathbf{u}_{i+1}, \qquad (6.13)$$
$$\hat{\mathbf{e}}_{i+3} = \mathbf{e}_i - 3G\Delta\mathbf{u}_{i+1}, \qquad (6.14)$$
$$\vdots$$
$$\hat{\mathbf{e}}_{i+p} = \mathbf{e}_i - pG\Delta\mathbf{u}_{i+1}. \qquad (6.15)$$

The optimisation problem becomes one associated with the cost function:

$$J = \sum_{j=1}^{p} \hat{\mathbf{e}}_{i+j}^T \hat{\mathbf{e}}_{i+j} + \gamma\Delta\mathbf{u}_{i+1}^T\Delta\mathbf{u}_{i+1}, \qquad (6.16)$$

where $\gamma = p\bar{\gamma}$. Substituting (6.12)–(6.15) into (6.16) yields

$$J = (\hat{\mathbf{e}}_{i+1}, \hat{\mathbf{e}}_{i+2}, ..., \hat{\mathbf{e}}_{i+p}) \begin{pmatrix} \hat{\mathbf{e}}_{i+1} \\ \hat{\mathbf{e}}_{i+2} \\ \vdots \\ \hat{\mathbf{e}}_{i+p} \end{pmatrix} + \gamma\Delta\mathbf{u}_{i+1}^T\Delta\mathbf{u}_{i+1}$$

$$= [\begin{pmatrix} I_{Nr} \\ I_{Nr} \\ \vdots \\ I_{Nr} \end{pmatrix} \mathbf{e}_i - \begin{pmatrix} I_{Nr} \\ 2I_{Nr} \\ \vdots \\ pI_{Nr} \end{pmatrix} G\Delta\mathbf{u}_{i+1}]^T [\begin{pmatrix} I_{Nr} \\ I_{Nr} \\ \vdots \\ I_{Nr} \end{pmatrix} \mathbf{e}_i$$

$$- \begin{pmatrix} I_{Nr} \\ 2I_{Nr} \\ \vdots \\ pI_{Nr} \end{pmatrix} G\Delta\mathbf{u}_{i+1}]$$

$$+ \gamma\Delta\mathbf{u}_{i+1}^T I_{Nm}\Delta\mathbf{u}_{i+1},$$

where I_{Nm} and I_{Nr} are the $Nm \times Nm$ and $Nr \times Nr$ unit matrices, respectively. Imposing the condition on the gradient $\frac{\partial J}{\partial\Delta\mathbf{u}_{i+1}} = 0$, the control action turns out to be

$$\Delta\mathbf{u}_{i+1} = (G^T F_1^T F_1 G + \gamma I_{Nm})^{-1} G^T F_1^T F_2 \mathbf{e}_i, \qquad (6.17)$$

where $F_1 = [I_{Nr}, 2I_{Nr}, ..., hI_{Nr}]^T$ and $F_2 = [I_{Nr}, I_{Nr}, ..., I_{Nr}]^T$. In this equation, it is noted that

$$F_1^T F_1 = [I_{Nr}, 2I_{Nr}, ..., pI_{Nr}] \begin{bmatrix} I_{Nr} \\ 2I_{Nr} \\ \vdots \\ pI_{Nr} \end{bmatrix} \tag{6.18}$$

$$= (1 + 2^2 + ... + p^2)I_{Nr} = aI_{Nr}, \tag{6.19}$$

where $a = \frac{1}{6}p(p+1)(2p+1)$. It is also noted that

$$F_1^T F_2 = [I_{Nr}, 2I_{Nr}, ..., pI_{Nr}] \begin{bmatrix} I_{Nr} \\ I_{Nr} \\ \vdots \\ I_{Nr} \end{bmatrix} \tag{6.20}$$

$$= (1 + 2 + ... + p)I_{Nr} = bI_{Nr}, \tag{6.21}$$

where $b = \frac{1}{2}p(1+p)$. Thus, (6.17) becomes

$$\mathbf{u}_{i+1} = \mathbf{u}_i + b(aG^T G + \gamma I_{Nm})^{-1}G^T \mathbf{e}_i, \tag{6.22}$$

where $\gamma > 0$.

Remark 6.1

The predictive ILC scheme has a feedforward structure, and the ith current input is generated by the previous data during the $(i-1)$th trial. The advantage of this algorithm is that the associated matrix has the same dimensions as that of the non-predictive ILC with quadratic criterion (Lee *et al.*, 2000), while keeping the basic predictive control features.

Remark 6.2

When the system disturbances and noise are significant, a similar observer algorithm to that of Lee *et al.* (2000), may be considered:

$$\mathbf{u}_{i+1} = \mathbf{u}_i + b(aG^T G + \gamma I_{Nm})^{-1}G^T \bar{\mathbf{e}}_{i/i}, \tag{6.23}$$

where $\bar{\mathbf{e}}_{i/i}$ is the estimate of $\mathbf{e}_i$. The following observer is used for estimating $\bar{\mathbf{e}}_{i/i}$:

$$\bar{\mathbf{e}}_{i/i-1} = \bar{\mathbf{e}}_{i-1/i-1} - G\Delta\mathbf{u}_i, \tag{6.24}$$

$$\bar{\mathbf{e}}_{i/i} = \bar{\mathbf{e}}_{i/i-1} + K(\mathbf{e}_i - \bar{\mathbf{e}}_{i/i-1}), \tag{6.25}$$

where K is the filter gain matrix which can be obtained through various means, such as pole placement and Kalman filtering techniques.

6.2.4 Convergence and Robustness of Algorithm

For the above control laws, the following convergence can be established

Theorem 6.1

For the system (6.6), with the assumption that $\mathbf{w}_{i+1} - \mathbf{w}_i = 0$, $x_{i+1}(0) - x_i(0) = 0$ and G is full row rank, given the desired trajectory $\mathbf{y}_d$ over the fixed time interval $[0, N]$, by using the learning control law (6.23), the tracking error converges to zero for $p \geq 1$ as $i \to \infty$.

Proof

Substituting (6.23) into (6.6) gives

$$\mathbf{e}_{i+1} = [I_{Nr} - bG(aG^TG + \gamma I_{Nm})^{-1}G^T]\mathbf{e}_i$$
$$= [I_{Nr} - \frac{b}{a}G(G^TG + \frac{\gamma}{a}I_{Nm})^{-1}G^T]\mathbf{e}_i.$$

Let

$$E = I_{Nr} - \frac{b}{a}G(G^TG + \frac{\gamma}{a}I_{Nm})^{-1}G^T, \qquad (6.26)$$

and using the matrix inversion lemma (Burl, 1999) yields

$$(G^TG + \frac{\gamma}{a}I_{Nm})^{-1} = \frac{a}{\gamma}I_{Nm} - \frac{a^2}{\gamma^2}G^T(\frac{a}{\gamma}GG^T + I_{Nr})^{-1}G$$
$$= \frac{a}{\gamma}[I_{Nm} - G^T(GG^T + \frac{\gamma}{a}I_{Nr})^{-1}G].$$

Applying this formula to (6.26) yields

$$E = I_{Nr} - \frac{b}{a}G(G^TG + \frac{\gamma}{a}I_{Nm})^{-1}G^T$$
$$= I_{Nr} - \frac{b}{\gamma}[GG^T - GG^T(GG^T + \frac{\gamma}{a}I_{Nr})^{-1}GG^T]$$
$$= I_{Nr} - \frac{b}{\gamma}[GG^T(GG^T + \frac{\gamma}{a}I_{Nr})^{-1}(GG^T + \frac{\gamma}{a}I_{Nr})$$

$$-GG^T(GG^T + \frac{\gamma}{a})^{-1}GG^T]$$
$$= I_{Nr} - \frac{b}{a}GG^T(GG^T + \frac{\gamma}{a}I_{Nr})^{-1}.$$

Since G is full row rank, GG^T is positive definite and non-singular (this is obtained directly from the singular value decomposition (SVD)). Thus,

$$E = I_{Nr} - \frac{b}{a}[(GG^T + \frac{\gamma}{a}I_{Nr})(GG^T)^{-1}]^{-1}$$
$$= I_{Nr} - \frac{b}{a}[I_{Nr} + \frac{\gamma}{a}(GG^T)^{-1}]^{-1}.$$

For the convergence analysis, it is necessary to know the eigenvalues of E:

$$\lambda_i(E) = \lambda_i\{I_{Nr} - \frac{b}{a}[I_{Nr} + \frac{\gamma}{a}(GG^T)^{-1}]^{-1}\} = 1$$
$$- \frac{b}{a}\lambda_i\{[I_{Nr} + \frac{\gamma}{a}(GG^T)^{-1}]^{-1}\}$$
$$= 1 - \frac{b}{a}\frac{1}{\{1 + \frac{\gamma}{a}\lambda_i[(GG^T)^{-1}]\}}.$$

Since GG^T is positive definite and non-singular, $0 < \frac{1}{1+\frac{\gamma}{a}\lambda_i[(GG^T)^{-1}]} < 1$. Also, since $0 < \frac{b}{a} = \frac{3}{2p+1} \leq 1$ for $p \geq 1$,

$$0 < \frac{b}{a}\frac{1}{\{1 + \frac{\gamma}{a}\lambda_i[(GG^T)^{-1}]\}} < 1. \qquad (6.27)$$

Note that E is a constant matrix. This implies that $|\lambda_i(E)| < 1$. The conclusion follows.

One advantage of the predictive ILC is the availability of tuning parameters, like the input weighting matrix γ and prediction horizon p, that can be used to enhance the robustness against model uncertainty. This is shown by considering a case where G contains the uncertainty. The error evolution equation for the true system is written as

$$\mathbf{e}_{i+1} = \mathbf{e}_i - G^{true}\Delta\mathbf{u}_{i+1} = \mathbf{e}_{i+1} - (G + \Delta G)\Delta u_{i+1}, \qquad (6.28)$$

where G is the nominal matrix, and ΔG is the perturbation matrix. Assume that

$$\| \Delta G \| \leq \varphi, \qquad (6.29)$$

where φ is a constant. The upper and lower bounds on φ are to be found such that if φ is within these bounds, the error convergence analysis still holds.

Since E is an asymptotically stable matrix as shown in the proof of Theorem 6.1, the following Lyapunov equation holds:

$$E^T PE - P = -I_{Nr}, \tag{6.30}$$

where P is a positive definite matrix. This equation will be used in the proof of the following theorem.

Theorem 6.2

For the system (6.28) with the assumption that $\mathbf{w}_{i+1} - \mathbf{w}_i = 0$, $x_{i+1}(0) - x_i(0) = 0$ and the nominal matrix G is full row rank, given the desired trajectory $\mathbf{y}_d$ over the fixed time interval $[0, N]$, by using the learning control law (6.23), then, the tracking error converges to zero for $p \geq 1$ as $i \to \infty$, if

$$\frac{-c - \sqrt{c^2 + d}}{d} < \varphi < \frac{-c + \sqrt{c^2 + d}}{d}, \tag{6.31}$$

where

$$c = \| E^T P \| \| (aG^T G + \gamma I_{Nm})^{-1} G^T \|, \tag{6.32}$$
$$d = \| (aG^T G + \gamma I_{Nm})^{-1} G^T \|^2 \| P \|. \tag{6.33}$$

Proof

Substituting (6.23) into (6.6) gives

$$\begin{aligned}
\mathbf{e}_{i+1} &= \mathbf{e}_i - (G + \Delta G)\Delta \mathbf{u}_{i+1} \\
&= [I_{Nr} - bG(aG^T G + \gamma I_{Nm})^{-1} G^T]\mathbf{e}_i \\
&\quad - b\Delta G(aG^T G + \gamma I_{Nm})^{-1} G^T \mathbf{e}_i \\
&= E\mathbf{e}_i - b\Delta G(aG^T G + \gamma I_{Nm})^{-1} G^T \mathbf{e}_i.
\end{aligned}$$

For simplicity, denote $H = (aG^T G + \gamma I_{Nm})^{-1} G^T$. Define the Lyapunov function $V_{i+1} = \mathbf{e}_{i+1}^T P \mathbf{e}_{i+1}$, i.e.,

$$\begin{aligned}
\Delta V_{i+1} &= V_{i+1} - V_i = \mathbf{e}_{i+1}^T P \mathbf{e}_{i+1} - \mathbf{e}_i^T P \mathbf{e}_i \\
&= (E\mathbf{e}_i - \Delta GH\mathbf{e}_i)^T P(E\mathbf{e}_i - \Delta GH\mathbf{e}_i) - \mathbf{e}_i^T P \mathbf{e}_i \\
&= \mathbf{e}_i^T (E^T PE - P)\mathbf{e}_i - 2\mathbf{e}_i^T E^T P\Delta GH\mathbf{e}_i + \mathbf{e}_i^T H^T \Delta G^T P\Delta GH\mathbf{e}_i \\
&\leq -\|\mathbf{e}_i\|^2 + 2\varphi \| E^T P \| \| H \| \| \mathbf{e}_i \|^2 + \varphi^2 \| H \|^2 \| P \| \| \mathbf{e}_i \|^2.
\end{aligned}$$

From (6.32) and (6.33),

$$\Delta V_{i+1} \leq (-1 + 2c\varphi + d\varphi^2)||\mathbf{e}_i||^2$$
$$= d(\varphi - \frac{-c + \sqrt{c^2 + d}}{d})(\varphi - \frac{-c - \sqrt{c^2 + d}}{d})||\mathbf{e}_i||^2. \tag{6.34}$$

Clearly, if $\frac{-c-\sqrt{c^2+d}}{d} < \varphi < \frac{-c+\sqrt{c^2+d}}{d}$, $d\varphi^2 + 2c\varphi - 1 < 0$. This together with $d > 0$, implies that convergence is achieved. This completes the proof.

In Theorem 6.1 and Theorem 6.2, convergence properties are established in the absence of the system measurement noise, disturbances, and initialisation errors. In practical applications, robustness of ILC algorithms against these uncertainties is an important issue to be addressed.

For convenience of presentation, the following notation is first defined.

$$\lambda_{G_0} = \lambda_{max}(G_0^T P G_0), \quad \lambda_P = \lambda_{max}(P), \quad P_{G_0} = ||G_0^T P||,$$
$$\lambda_{G_0 E} = \lambda_{max}[(G_0^T P E)(E^T P G_0)], \quad \lambda_E = \lambda_{max}[(PE)(E^T P)].$$

Theorem 6.3

For the system (6.6) with the assumptions that $||\mathbf{w}_i - \mathbf{w}_{i-1}|| \leq b_w, ||x_i(0) - x_{i-1}(0)|| \leq b_{x0}$ and G is full row rank, given the desired trajectory $\mathbf{y}_d$ over the fixed time interval $[0, N]$, by using the learning control law (6.23), then, the tracking error converges to the following bound for $p \geq 1$ as $i \to \infty$,

$$\lim_{i \to \infty} ||\mathbf{e}_i|| \leq \sqrt{\frac{g(b_{x0}, b_w)}{(1 - \rho)\lambda_{min}(P)}}, \tag{6.35}$$

where $g(b_{x0}, b_w)$ is constant proportional to constants b_{x0}, and b_w. Moreover, $\lim_{i \to \infty} ||\mathbf{e}_i|| = 0$ if $b_{x0} = b_w = 0$.

Proof

The error model is derived according to the current conditions.

$$\mathbf{e}_{i+1} = [I_{Nr} - bG(aG^T G + \gamma I_{Nm})^{-1} G^T]\mathbf{e}_i - G_0 \Delta x_{i+1}(0) - \Delta \mathbf{w}_{i+1}$$
$$= E\mathbf{e}_i - G_0 \Delta x_{i+1}(0) - \Delta \mathbf{w}_{i+1}.$$

Similar to the proof in Theorem 6.2, the Lyapunov function can be selected as

$$V_i = \mathbf{e}_i^T P \mathbf{e}_i.$$

Then, it can be shown that

$$\Delta V_{i+1} = V_{i+1} - V_i = \mathbf{e}_{i+1}^T P \mathbf{e}_{i+1} - \mathbf{e}_i^T P \mathbf{e}_i$$

$$= [E\mathbf{e}_i - G_0 \Delta x_{i+1}(0) - \Delta \mathbf{w}_{i+1}]^T P[E\mathbf{e}_i - G_0 \Delta x_{i+1}(0)$$
$$-\Delta \mathbf{w}_{i+1}] - \mathbf{e}_i^T P \mathbf{e}_i$$

$$= \mathbf{e}_i^T (E^T P E - P)\mathbf{e}_i - 2\mathbf{e}_i^T E^T P G_0 \Delta x_{i+1}(0) - 2\mathbf{e}_i^T E^T P \Delta \mathbf{w}_{i+1}$$
$$+2\Delta x_{i+1}^T(0)G_0^T P \Delta \mathbf{w}_{i+1}$$
$$+\Delta x_{i+1}^T(0)G_0^T P G_0 \Delta x_{i+1}(0) + \Delta \mathbf{w}_{i+1}^T P \Delta \mathbf{w}_{i+1}$$

$$= -\mathbf{e}_i^T \mathbf{e}_i - 2\mathbf{e}_i^T E^T P G_0 \Delta x_{i+1}(0) - 2\mathbf{e}_i^T E^T P \Delta \mathbf{w}_{i+1}$$
$$+2\Delta x_{i+1}^T(0)G_0^T P \Delta \mathbf{w}_{i+1}$$
$$+\Delta x_{i+1}^T(0)G_0^T P G_0 \Delta x_{i+1}(0) + \Delta \mathbf{w}_{i+1}^T P \Delta \mathbf{w}_{i+1}. \qquad (6.36)$$

Using $-2\alpha^T \beta \leq \eta \alpha^T \alpha + \frac{1}{\eta}\beta^T \beta$ where η is an arbitrarily positive constant, the following inequalities hold:

$$-2\mathbf{e}_i^T E^T P G_0 \Delta x_{i+1}(0) \leq \eta \mathbf{e}_i^T \mathbf{e}_i$$
$$+\frac{1}{\eta}\Delta x_{i+1}^T(0)(G_0^T P E)(E^T P G_0)\Delta x_{i+1}(0)$$
$$\leq \eta \|\mathbf{e}_i\|^2 + \frac{1}{\eta}\lambda_{max}[(G_0^T P E)(E^T P G_0)]b_{x0}^2,$$

$$-2\mathbf{e}_i^T E^T P \Delta \mathbf{w}_{i+1} \leq \eta \|\mathbf{e}_i\|^2 + \frac{1}{\eta}\lambda_{max}[(P E)(E^T P)]b_w^2.$$

Note that

$$\Delta x_{i+1}^T(0)G_0^T P G_0 \Delta x_{i+1}(0) \leq \lambda_{max}(G_0^T P G_0)b_{x0}^2,$$
$$\Delta \mathbf{w}_{i+1}^T P \Delta \mathbf{w}_{i+1} \leq \lambda_{max}(P)b_w^2,$$
$$2\Delta x_{i+1}(0)^T G_0^T P \Delta \mathbf{w}_{i+1} \leq 2\|G_0^T P\|b_{x0}b_w.$$

According to the definition of the sup-norm,

$$\mathbf{e}_i^T \mathbf{e}_i \geq \|\mathbf{e}_i\|^2.$$

This implies that $-\mathbf{e}_i^T \mathbf{e}_i \leq -\|\mathbf{e}_i\|^2$. Then, it can be shown that

$$\Delta V_{i+1} \leq -(1 - 2\eta)\|\mathbf{e}_i\|^2 + g(b_{x0}, b_w), \qquad (6.37)$$

where $g(b_{x0}, b_w) = (\lambda_{G_0} + \frac{1}{\eta}\lambda_{G_0 E})b_{x0}^2 + 2P_{G_0}b_{x0}b_w + (\lambda_P + \frac{1}{\eta}\lambda_E)b_w^2$.

Since $\lambda_{min}(P)\|\mathbf{e}_i\|^2 \leq V_i \leq \lambda_{max}(P)\|\mathbf{e}_i\|^2$, it follows that

$$V_{i+1} - V_i \le -\frac{1-2\eta}{\lambda_{max}(P)}V_i + g(b_{x0}, b_w), \tag{6.38}$$

$$V_{i+1} \le [1 - \frac{1-2\eta}{\lambda_{max}(P)}]V_i + g(b_{x0}, b_w) = \rho V_i + g(b_{x0}, b_w), \tag{6.39}$$

where the value of η is chosen as:

$$\frac{1 - \lambda_{max}(P)}{2} < \eta < 1/2 \ \ if \ \lambda_{max}(P) < 1, \tag{6.40}$$

$$0 < \eta < 1/2 \ \ if \ \lambda_{max}(P) \ge 1, \tag{6.41}$$

such that $0 < \rho < 1$. Finally,

$$V_i \le \rho^i V_0 + \frac{1 - \rho^i}{1 - \rho} g(b_{x0}, b_w), \tag{6.42}$$

$$\lim_{i \to \infty} V_i \le \frac{g(b_{x0}, b_w)}{1 - \rho}, \tag{6.43}$$

$$\lim_{i \to \infty} \|e_i\| \le \sqrt{\frac{g(b_{x0}, b_w)}{(1 - \rho)\lambda_{min}(P)}}. \tag{6.44}$$

Hence, $\lim_{i \to \infty} \|e_i\| = 0$ if $b_{x0}, b_w \to 0$. This completes the proof.

6.2.5 Simulation

Consider the injection moulding example first described in Chapter 4. A discrete-system state equation in the time domain may be obtained directly from (4.59) with zero-order hold and the sampling time $T = 0.005$s, given by

$$x(k + 1) = Ax(k) + Bu(k) \tag{6.45}$$

$$y(k) = Cx(k), \ \ k \in [0, N], \tag{6.46}$$

where

$$A = \begin{bmatrix} 0.1065 & 66.5118 & -6.6885 \times 10^4 & -1.0237 \times 10^7 \\ 5.0154 \times 10^{-5} & 0.2083 & 194.1345 & 2.9474 \times 10^4 \\ -1.4440 \times 10^{-7} & -2.4283 \times 10^{-4} & -0.1591 & -83.2888 \\ 4.0805 \times 10^{-10} & 6.8353 \times 10^{-7} & 7.9549 \times 10^{-4} & 0.6248 \end{bmatrix}$$

$$B = \begin{bmatrix} 5.0154 \times 10^{-5} \\ -1.444 \times 10^{-7} \\ 4.0805 \times 10^{-10} \\ 1.8381 \times 10^{-12} \end{bmatrix}$$

$$C = [0 \ 0 \ 0 \ 2.144 \times 10^{11}].$$

$y(k)$ is a measurable voltage signal proportional to the ram velocity, $u(k)$ is the controlling variable which is a signal proportional to the current input to the servovalve, and k is a discrete time over $[0, N]$.

Since injection moulding is a cyclical process, it is more attractive to use a learning controller. In this section, a predictive learning controller is used for the cycle-to-cycle control of the injection moulding process. First, to illustrate the performance of the controller, simulations with prediction horizon $p = 4$ and $r = 0.05$ are shown in Fig. 6.6 at the 1st, 2nd, 3rd, and 10th cycles. It is observed that the predictive ILC scheme yields good set-point tracking performance as the cycle number increases.

To test the robustness of the algorithm, assume that there is a modelling error between the identified and actual models. The learning controller is designed based on the nominal model derived from

$$G_m(s) = \frac{2 \times 10^{11}}{(s + 100)(s + 1100)[(s + 383)^2 + 1135^2]}, \qquad (6.47)$$

which has a large model uncertainty compared to the actual model (4.59). Figure 6.7 shows the control result of the predictive learning algorithm. Clearly, the control can achieve satisfactory tracking performance even with perturbation of the model parameters.

To further test the robustness of the algorithm, measurement noise is introduced into the simulation. Now, the process uncertainties include model error and measurement noise. The control performance is shown in Fig. 6.8. Next, a repetitive form of disturbance described by $\sin(0.0314t)$ is added. Figure 6.5 shows the control performance. It can be observed that good convergence and robustness properties are still achieved as the iteration number increases.

6.3 Predictive Iterative Learning Control without Package Information

To accommodate the advantages of predictive control, the ILC of Section 6.2 has been combined with predictive control in an appropriate way. However, these results have to use the package information matrix which includes the whole time interval for an interaction. It has this disadvantage which becomes clearly significant with increasing control precision requirements. The package information matrix has extensive memory requirements. This can lead to

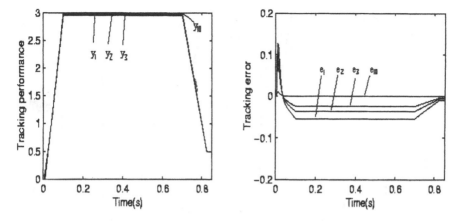

Fig. 6.2. Tracking performance of the controller: no uncertainty is considered

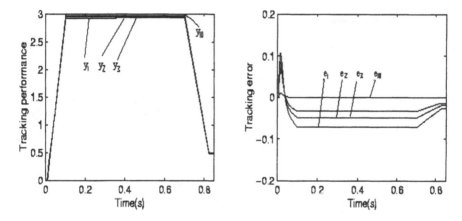

Fig. 6.3. Tracking performance of the controller: model error is considered

numerical difficulties in computing the ILC control law. For example, when a control system adopts a small sampling time (*e.g.*, 0.001s), and long interval (*e.g.*, 1s), the package information matrix becomes huge (1000 × 1000). This difficulty is especially significant for robotic manipulators, where a small sampling time is usually necessary in order to compensate errors to within an acceptable precise threshold.

In this section, a predictive learning control is developed for repetitive processes, which does not require package information. The basic idea of the control is to improve the control signal for the present operation repetition by feeding back the control error from previous and future iterations. The predictor is designed based on the repetition number. The error convergence is analysed rigorously, and the analysis will be provided in this section. The

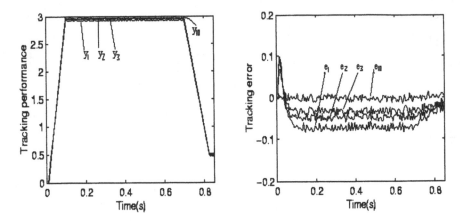

Fig. 6.4. Tracking performance of the controller: model error and measurement noise are considered

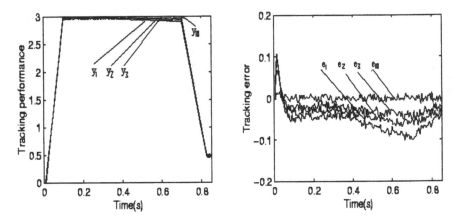

Fig. 6.5. Tracking performance of the controller: model error, measurement noise and repetitive disturbance are considered

injection moulding example is used again to demonstrate the effectiveness of the algorithm.

6.3.1 Problem Statement

Consider a repetitive linear discrete time-invariant system with uncertainty and disturbance as follows:

$$
\begin{aligned}
x_i(t+1) &= Ax_i(t) + Bu_i(t) + w_i(t), \\
y_i(t) &= Hx_i(t), \quad t \in [0, N],
\end{aligned}
\tag{6.48}
$$

where i denotes the ith repetitive operation of the system; $x_i(t) \in R^n, u_i(t) \in R^m$, and $y_i(t) \in R^r$ are the state, control input, and output of the system, respectively; $w_i(t)$ is uncertainty or disturbance; $t \in [0, N]$ is the time; and A, B, and H are matrices with appropriate dimensions.

The problem is stated as follows: an update mechanism is to be formulated for the input trajectory of a new iteration based on the information from the previous iteration, so that the controlled output converges to the desired reference over the time horizon $[0, N]$.

6.3.2 Error Propagation

To determine how the error evolves with repetitions, solving the equation (6.48) produces

$$x_i(1) = Ax_i(0) + Bu_i(0) + w_i(0),$$
$$x_i(2) = A^2 x_i(0) + ABu_i(0) + Bu_i(1) + Aw_i(0) + w_i(1),$$
$$\vdots$$
$$x_i(t) = A^t x_i(0) + A^{t-1} Bu_i(0) + ... + Bu_i(t-1) + A^{t-1} w_i(0)$$
$$+ ... + w_i(t-1).$$

Let

$$W_i(t-1) = A^{t-1} w_i(0) + ... + w_i(t-1). \tag{6.49}$$

Thus, the solution of (6.48) becomes

$$x_i(t) = A^t x_i(0) + A^{t-1} Bu_i(0) + ... + Bu_i(t-1) + W_i(t-1). \tag{6.50}$$

The tracking error $e_i(t+1)$ at the ith repetition is given by $e_i(t+1) = y_d(t+1) - y_i(t+1)$. From this definition, one can reach the following output error equation:

$$\begin{aligned} e_{i+1}(t+1) &= y_d(t+1) - y_{i+1}(t+1) \\ &= y_d(t+1) - y_i(t+1) - (y_{i+1}(t+1) - y_i(t+1)) \\ &= e_i(t+1) - (y_{i+1}(t+1) - y_i(t+1)). \end{aligned} \tag{6.51}$$

It can be directly shown that $y_{i+1}(t+1) - y_i(t+1)$.

$$y_{i+1}(t+1) - y_i(t+1) = HA^{t+1}(x_{i+1}(0) - x_i(0))$$
$$+CA^tB(u_{i+1}(0) - u_i(0)) + \dots$$
$$+HAB(u_{i+1}(t-1) - u_i(t-1))$$
$$+HB(u_{i+1}(t) - u_i(t))$$
$$+H(W_{i+1}(t) - W_i(t)).$$

Substituting the above equation into (6.51) yields

$$
\begin{aligned}
e_{i+1}(t+1) &= e_i(t+1) - HA^{t+1}\Delta x_{i+1}(0) - HA^tB\Delta u_{i+1}(0) - \dots \\
&\quad - HAB\Delta u_{i+1}(t-1) - HB\Delta u_{i+1}(t) - HAW_{i+1}(t) \\
&= e_i(t+1) - HA^{t+1}\Delta x_{i+1}(0) - G^T(t)\Delta U_{i+1} \\
&\quad - H\Delta W_{i+1}(t),
\end{aligned}
\tag{6.52}
$$

where $\Delta x_{i+1}(0) = x_{i+1}(0) - x_i(0), \Delta U_{i+1}(t-1) = U_{i+1}(t-1) - U_i(t-1), \Delta W_{i+1}(t-1) = W_{i+1}(t-1) - W_i(t-1)$, and

$$G^T(t) = [HA^tB, HA^{t-1}B, \dots, HB], \tag{6.53}$$
$$\Delta U_i^T = [\Delta u_i(0), \Delta u_i(1), \dots, \Delta u_i(t)]. \tag{6.54}$$

6.3.3 Predictor Construction

Once the state space model is available, the subsequent steps for predictor construction are straightforward. For this, the structure of the model (6.52) is used in formulating the predictive controller. In order to do this, define a state prediction model of the form

$$\hat{e}_{i+1}(t+1) = \hat{e}_i(t+1) - G^T(t)\Delta U_{i+1}, \tag{6.55}$$

where $\hat{e}_{i+1}(t+1)$ denotes the error predicted at instant i. This model is redefined at each repetition i from the actual error, that is,

$$\hat{e}_i(t+1) = e_i(t+1). \tag{6.56}$$

As in the preceding section, $\Delta x_i(0)$ and the disturbance are not in the prediction model. By applying the equation (6.55) recursively, it follows that

$$\hat{e}_{i+2}(t+1) = e_i(t+1) - G^T(t)\Delta U_{i+2} - G^T(t)\Delta U_{i+1}, \tag{6.57}$$
$$\hat{e}_{i+3}(t+1) = e_i(t+1) - G^T(t)[\Delta U_{i+3} + \Delta U_{i+2} + \Delta U_{i+1}], \tag{6.58}$$

$$\vdots$$

$$\hat{e}_{i+m}(t+1) = e_i(t+1) - G^T(t)[\Delta U_{i+m} + \Delta U_{i+m-1} + ...$$
$$+ \Delta U_{i+1}]. \tag{6.59}$$

The above equations may be rewritten in composite form:

$$\mathcal{E} = \mathcal{F}e_i(t) - \mathcal{G}\mathcal{U}, \tag{6.60}$$

where

$$\mathcal{E} = \begin{bmatrix} \hat{e}_{i+1}(t+1) \\ \hat{e}_{i+2}(t+1) \\ \vdots \\ \hat{e}_{i+m}(t+1) \end{bmatrix}, \quad \mathcal{F} = [1, 1, ..., 1]^T,$$

$$\mathcal{G} = \begin{bmatrix} G^T(t) & 0 & ... & 0 \\ G^T(t) & G^T(t) & ... & 0 \\ & \vdots & & \\ G^T(t) & G^T(t) & ... & G^T(t) \end{bmatrix}, \quad \mathcal{U} = \begin{bmatrix} \Delta U_{i+1} \\ \Delta U_{i+2} \\ \vdots \\ \Delta U_{i+m} \end{bmatrix}.$$

The control sequence is still made constant over the prediction interval, $i.e.$,

$$\Delta U_{i+1} = \Delta U_{i+2} = ... = \Delta U_{i+m}. \tag{6.61}$$

(6.60) becomes

$$\mathcal{E} = \mathcal{F}e_i(t+1) - \mathcal{G}_1 \Delta U_{i+1}, \tag{6.62}$$

where

$$\mathcal{G}_1 = \begin{bmatrix} G^T(t) \\ 2G^T(t) \\ \vdots \\ mG^T(t) \end{bmatrix}. \tag{6.63}$$

For the time t, since $\Delta u_{i+1}(0), ..., \Delta u_{i+1}(t-1)$ are available, (6.62) can be rewritten as

$$\mathcal{E} = \mathcal{F}e_i(t+1) - G_1^T(k-1)\Delta \bar{U}_{i+1}(t-1) - G_2^T \Delta u_{i+1}(t), \tag{6.64}$$

where

$$G_1^T(t-1) = \begin{bmatrix} [CA^tB, CA^{t-1}B, ..., CAB] \\ 2[CA^tB, CA^{t-1}B, ..., CAB] \\ \vdots \\ m[CA^tB, CA^{t-1}B, ..., CAB] \end{bmatrix}, \qquad (6.65)$$

$$\Delta\bar{U}_{i+1}^T(t-1) = [\Delta u_{i+1}(0), \Delta u_{i+1}(1), ..., \Delta u_{i+1}(t-1)], \qquad (6.66)$$

$$G_2^T = \begin{bmatrix} CB \\ 2CB \\ \vdots \\ mCB \end{bmatrix}. \qquad (6.67)$$

6.3.4 Derivation of Predictive Learning Algorithm

Consider a linear quadratic performance index written in the following form:

$$J = \mathcal{E}^T Q \mathcal{E} + \Delta u_{i+1}^T(t) R \Delta u_{i+1}(t), \qquad (6.68)$$

where Q, R are positive definite matrices. The following update law can easily be obtained by substituting (6.62) into (6.68)

$$\Delta u_{i+1}(t) = (G_2^T Q G_2 + R)^{-1} G_2^T Q$$
$$\times [\mathcal{F}e_i(t+1) - G_1^T(t-1)\Delta\bar{U}_{i+1}(t-1)]. \qquad (6.69)$$

Note that the learning control law (6.69) differs from that of the preceding section, since the previous results are based on the package information matrix. The learning law at time t utilises $e_i(t+1)$ at the previous repetition and part values $\Delta u_{i+1}(0), \Delta u_{i+1}(1), ..., \Delta u_{i+1}(t-1)$ at previous times $0, 1, ..., t-1$.

6.3.5 Convergence of Predictive Learning Algorithm

For the control (6.69), the following convergence properties can be established.

Theorem 6.4

Consider the system (6.48) and the assumptions $W_{i+1} - W_i(k) = 0$, $x_{i+1}(0) - x_i(0) = 0$ and CB being full row rank. Given the desired trajectory y_d over the fixed time interval $[0, N]$, by using the learning control law (6.69), the tracking error converges to zero as $i \to \infty$, if

$$\rho = \| I - CB(G_2^T QG_2 + R)^{-1}G_2^T Q\mathcal{F} \| < 1, \tag{6.70}$$

where I is the unit matrix.

Proof

Let $V = (G_2^T QG_2 + R)^{-1}G_2^T Q$. Applying the control law (6.69) to (6.52) yields

$$\Delta u_{i+1}(0) = VFe_i(1),$$
$$e_{i+1}(1) = e_i(1) - CB\Delta u_{i+1}(0) = (I - CBVF)e_i(1).$$

Taking norms yields

$$\|e_{i+1}(1)\| \le \rho\|e_i(1)\|. \tag{6.71}$$

Furthermore,

$$\Delta u_{i+1}(1) = VFe_i(2) - VG_1^T(0)\Delta u_{i+1}(0). \tag{6.72}$$

Substituting $\Delta u_{i+1}(0)$ into the above control yields

$$\begin{aligned}
\Delta u_{i+1}(1) &= VFe_i(2) - VG_1^T(0)VFe_i(1)\\
e_{i+1}(2) &= e_i(2) - CAB\Delta u_{i+1}(0) - CB\Delta u_{i+1}(1)\\
&= e_i(2) - CABVFe_i(1) - CBVFe_i(2)\\
&\quad + CBVG_1^T(0)VFe_i(1)\\
&= (I - CBVF)e_i(2) + (CBVG_1^T(0)VF - CABVF)e_i(1).
\end{aligned}$$

Taking norms yields

$$\|e_{i+1}(2)\| \le \rho\|e_i(2)\| + h_{2,1}\|e_i(1)\|, \tag{6.73}$$

where $h_{2,1} = \|CBVG_1^T(0)VF - CABVF\|$.

For $k = 3$ at the $(i+1)$th repetition,

$$\begin{aligned}
e_{i+1}(3) &= e_i(3) - CA^2B\Delta u_{i+1}(0) - CAB\Delta u_{i+1}(1) - CB\Delta u_{i+1}(2)\\
&= (I - CBVF)e_i(3) - CA^2B\Delta u_{i+1}(0) - CAB\Delta u_{i+1}(1)\\
&\quad + CBVG_1^T(1)\begin{bmatrix} \Delta u_{i+1}(0)\\ \Delta u_{i+1}(1) \end{bmatrix}\\
&= (I - CBVF)e_i(3) + (CBVG_{11}^T(1) - CA^2B)\Delta u_{i+1}(0)\\
&\quad + (CBVG_{12}^T(1) - CAB)\Delta u_{i+1}(1), \tag{6.74}
\end{aligned}$$

where

$$G_{11}^T(1) = \begin{bmatrix} CA^2B \\ 2CA^2B \\ \vdots \\ mCA^2B \end{bmatrix}, \qquad (6.75)$$

$$G_{12}^T(1) = \begin{bmatrix} CAB \\ 2CAB \\ \vdots \\ mCAB \end{bmatrix}. \qquad (6.76)$$

Substituting $\Delta u_{i+1}(0), \Delta u_{i+1}(1)$ into (6.76) and taking norms yields

$$||e_{i+1}(3)|| \le \rho||e_i(3)|| + h_{3,1}||e_i(1)|| + h_{3,2}||e_i(2)||, \qquad (6.77)$$

where

$$h_{3,1} = ||[CBVG_{11}^T(1) - CA^2B - CBVG_{12}^T(1)VG_1^T(0) \\ + CABVG_1^T(0)]VF||,$$
$$h_{3,2} = ||(CBVG_{12}^T(1) - CAB)VF||.$$

Similarly,

$$||e_{i+1}(k+1)|| \le \rho||e_i(k+1)|| + h_{k+1,1}||e_i(1)|| + h_{k+1,2}||e_i(2)|| + \cdots \\ + h_{k+1,k}||e_i(k)||.$$

The above inequalities can be written in composite form:

$$\begin{bmatrix} ||e_{i+1}(1)|| \\ ||e_{i+1}(2)|| \\ ||e_{i+1}(3)|| \\ \vdots \\ ||e_{i+1}(k+1)|| \end{bmatrix} \le \begin{bmatrix} \rho & 0 & 0 & \cdots & 0 \\ h_{2,1} & \rho & 0 & \cdots & 0 \\ h_{3,1} & h_{3,2} & \rho & \cdots & 0 \\ \vdots & \vdots & \vdots & \vdots & \vdots \\ h_{k+1,1} & h_{k+1,2} & h_{k+1,3} & \cdots & \rho \end{bmatrix} \begin{bmatrix} ||e_i(1)|| \\ ||e_i(2)|| \\ ||e_i(3)|| \\ \vdots \\ ||e_i(k+1)|| \end{bmatrix}.$$

Since $\rho < 1$, the conclusion follows.

6.3.6 Simulation

In this section, the same injection moulding as in the preceding section is used again. First, to illustrate the performance of the controller, simulations with $m = 0$ are carried out. This may be referred to as a conventional ILC scheme. In this case, one may choose $Q = 1$ and $R = 0.5$. Figure 6.6 shows the control performances at the 1st, 2nd, 3rd, and 10th cycles. The results are not satisfactory. When the predictive horizon is used in the ILC scheme (it is chosen as $m = 4$), Fig. 6.7 shows the control results for the same cycle number. In comparison with the case $m = 0$, it is observed that the predictive ILC scheme yields good set-point tracking performance. In addition, the control weighting R is also used to improve tracking performance. For example, if $R = 0.1$, the tracking error is further suppressed as shown in Fig. 6.8.

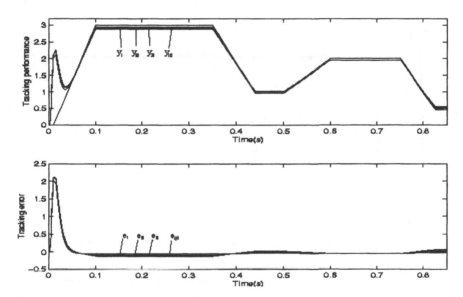

Fig. 6.6. Control performance of ILC with m=0

6.4 Non-linear Predictive Iterative Learning Control

In the previous sections, PILC has been investigated based on linear systems. This section will study methods to account for non-linearities without resorting to on-line parameter estimation as required in self-tuning control. A non-linear PILC-based algorithm is presented which operates along the time axis. In order to take into account system non-linearity, a neural-network-based technique is used for developing non-linear dynamic models from empirical data. The controller requires the extraction of a linear model from the neural network model. For the PILC dynamic linearisation at each time instant is

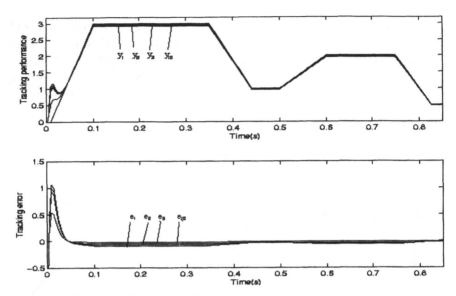

Fig. 6.7. Control performance with $m = 4$

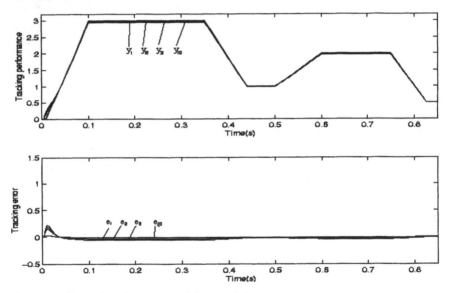

Fig. 6.8. Control performance with $R = 0.1$

done that exploits the special structure of the local linear models. As a major benefit of the linearisation, linear PILC can be used.

6.4.1 System Model

A large class of single-input single-output (SISO) non-linear dynamic systemes can be described in the discrete-time domain by:

$$y_i(t+1) = f(y_i(t), ..., y_i(t - n_y), u_i(t), ..., u_i(t - n_u)), \qquad (6.78)$$

where t is the time axis, i is the trial number, and $y_i(t)$ and $u_i(t)$ are the system output and input at the ith trial, respectively. The unknown function $f(\cdot)$ has to be approximated from measured data. In this section, a neural network (NN) model is applied to this task.

Let $f(\chi)$ be a smooth function from R^n to R. Then, given a compact $S \in R$ and a positive number ϵ_M, there exists a NN $\hat{f}(\chi)$ such that

$$f(\chi) = \hat{f}(\chi; W) + \epsilon, \qquad (6.79)$$

where W is the network weightings and $||\epsilon|| < \epsilon_M$ for all $\chi \in S$.

χ is a regression vector given by

$$\chi = [y(t), ..., y(t - n_y), u(t), ..., u(t - n_u)]^T. \qquad (6.80)$$

With the measured signals, parameters in W need to be defined to obtain the global system model. A commonly used weight tuning algorithm is the gradient algorithm based on a backpropagated error, where the NN is trained off-line to match specified exemplary pairs (χ_d, Υ_d), with χ_d being the ideal NN input that yields the desired NN output Υ_d. The learning procedure aims at driving the total error to near zero via suitable adjustments of the learning parameters. This essentially constitutes a minimisation problem which the gradient techniques attempts to solve.

6.4.2 Dynamic Linearisation

As in Chapter 3, the objective of dynamic linearisation is to calculate the time-invariant parameters $a_i(t)$ and $b_i(t)$ of the linear transfer function

$$G(z, t) = \frac{b_1 z^{-1} + b_2 z^{-2} + ... + b_n z^{-n}}{1 + a_1 z^{-1} + a_2 z^{-2} + ... + a_n z^{-n}} \qquad (6.81)$$

for arbitrary system states.

The parameters are obtained by first-order Taylor series approximation of the non-linear model:

$$a_k(t) = -\frac{\partial y(t)}{\partial y(t-i)}|_{x=x(t)}, \tag{6.82}$$

$$b_k(t) = -\frac{\partial y(t)}{\partial u(t-i)}|_{x=x(t)}. \tag{6.83}$$

The detailed derivatives in the NN structure can be found in Chapter 3.

The model (6.81) can also be written in the form

$$x_i(t+1) = Ax_i(t) + Bu_i(t), \tag{6.84}$$

$$y_i(t) = Hx_i(t), \tag{6.85}$$

where

$$A = \begin{bmatrix} -a_1 & 1 \dots 0 \\ -a_2 & 0 \dots 0 \\ \vdots & \vdots \dots \vdots \\ -a_{n-1} & 0 \dots 1 \\ -a_n & 0 \dots 0 \end{bmatrix}, \tag{6.86}$$

$$B = \begin{bmatrix} b_1 \\ b_2 \\ \vdots \\ b_{n-1} \\ b_n \end{bmatrix}, \tag{6.87}$$

$$H = [1\ 0\ ...0]. \tag{6.88}$$

6.4.3 Predictive Learning Control Scheme

The structure of the model (6.84) is used for formulating predictive controllers. First define a state prediction model of the form

$$\hat{x}_i(t+j/t) = A\hat{x}_i(t+j-1/t) + Bu_i(t+j-1/t), \; j = 1, 2, ..., p, \tag{6.89}$$

where $\hat{x}_i(t+j/t)$ denotes the state vector prediction at instant t for instant $t+j$ and $u_i(\cdot/t)$ denotes the sequence of control vectors on the prediction interval. This model is redefined at each sampling instant t from the actual state vector and the controls previously applied,

$$\hat{x}_i(t/t) = x_i(t); \quad j = 1, 2, ..., h. \tag{6.90}$$

Applying (6.89) recursively to the initial conditions gives

$$\left.\begin{array}{rl}
\hat{y}_i(t+1/t) = & HAx_i(t) + HBu_i(t/t), \\
\hat{y}_i(t+2/t) = & HA^2x_i(t) + HABu_i(t/t) + HBu_i(t+1/t), \\
& \vdots \\
\hat{y}_i(t+p_1/t) = & HA^{p_1}x_i(t) + HA^{p_1-1}Bu_i(t/t) + ... \\
& + HA^{p_1-p_2-1}Bu_i(t+p_2/t).
\end{array}\right\} \tag{6.91}$$

A composite form of the equation (6.91) can be written as follows:

$$Y_i = Gx_i(t) + F_1 U_i + F_1 U_{i-1}, \tag{6.92}$$

where

$$Y_i = [\hat{y}_i(t+1/t) \ \hat{y}_i(t+2/t)...\hat{y}_i(t+p_1/t)]^T,$$

$$G = \begin{bmatrix} HA \\ HA^2 \\ \vdots \\ HA^{p_1} \end{bmatrix},$$

$$F_1 = \begin{bmatrix} HB & 0 & ... & 0 \\ HAB & HB & ... & 0 \\ \vdots & \vdots & ... & \vdots \\ HA^{p_1-1}B & HA^{p_1-2}B & ... & HA^{p_1-p_2-1}B \end{bmatrix},$$

$$U_i = [u_i(t/t), u_i(t+1/t), ..., u_i(t+p_2/t)]^T, \tag{6.93}$$

where

$$\Delta U_i = U_i - U_{i-1}. \tag{6.94}$$

Note that incremental control is along the trial number. The performance index is given by

$$J = (Y_d - Y_i)^T(Y_d - Y_i) + r\Delta U_i^T \Delta U_i, \tag{6.95}$$

where $Y_d = [y_d(t+1/t), y_d(t+2/t), ..., y_d(t+p_1/t)]^T.$

The unconstrained optimal control is given by

$$\Delta U_i = (F_1^T F_1 + rI)^{-1} F_1^T (Y_d - Gx_i(t) - F_1 U_{i-1}). \tag{6.96}$$

The mathematical formulation of the constrained control problem for (6.95) is as follows:

$$min\ J = (Y_d - Y_i)^T (Y_d - Y_i) + r\Delta U_i^T \Delta U_i, \tag{6.97}$$

subject to

$$Y_{min} \le Y_i \le Y_{max},$$
$$U^{low} \le U_i \le U^{hi},$$
$$\Delta U^{low} \le U_i \le \Delta U^{hi}.$$

In order to solve the quadratic optimisation problem of the equation (6.97) subject to the constraints, the objective function and constraints are required to be defined as functions of the manipulating variables.

Using (6.96) and after removing the constant terms, the complete optimisation problem can be written in the following quadratic form:

$$min\ J = (Y_d - Y_i)^T (Y_d - Y_i) + r\Delta U_i^T \Delta U_i \tag{6.98}$$

subject to

$$\begin{bmatrix} F_1 \\ -F_1 \\ I \\ -I \\ I \\ -I \end{bmatrix} \Delta U_i \le \begin{bmatrix} Y_{max} - Gx_i(t) - F_1 U_{i-1} \\ -Y_{min} + Gx_i(t) + F_1 U_{i-1} \\ \Delta U^{hi} \\ -\Delta U^{low} \\ U^{hi} - U_{i-1} \\ -U^{low} + U_{i-1} \end{bmatrix}. \tag{6.99}$$

The quadratic programming algorithm available in MATLAB/SIMULINK Toolbox can be used for optimisation.

Remark 6.3

The present result differs from that of the conventional GPC given in Chapter 3. In the PILC scheme, the prediction capability is only exhibited along the time axis, while the control learning is along trial number. In the GPC scheme, both prediction and control ocurs along the time axis.

CHAPTER 7

ADAPTIVE PREDICTIVE CONTROL OF A CLASS OF SISO NON-LINEAR SYSTEMS

In the preceding chapters, predictive controller design techniques, given complete knowledge of the system parameters, have been presented. In practice, however, such knowledge is more often than not unavailable, introducing some uncertainty in the dynamical description of the system. This uncertainty can be in the form of a general dynamical operator (dynamic uncertainty) and/or in the form of parametric uncertainty, that is an error in the parameters of the state space representation of the nominal system. Some robustness properties with respect to modelling error and parametric uncertainty have been discussed in preceding chapters. However, they may fail to ensure closed-loop stability if the uncertainty is large enough. This problem becomes more pronounced in cases where the system parameters may vary in an unknown fashion over a wide range. In this chapter, the objective is to design an adaptive predictive controller without requiring complete knowledge of the system parameters. In this scheme, the system is first identified on-line and then a predictive controller is designed. Based on pseudo-partial-derivatives (PPD) derived on-line from the input and output information of the system, the system is identified on-line recursively using a modified least squares identification algorithm. The generalised predictive control (GPC) technique developed in Clarke *et al.* (1987) is applied to design the adaptive predictive control of the system. The stability of the adaptive scheme in the presence of uncertainty is analysed and proven. Simulation and real-time experiment examples are provided to illustrate the effectiveness of the proposed method when applied to real control problems.

7.1 Design of Adaptive Predictive Controller

As shown in the preceding chapters, if a good model is available *a priori* to describe the dynamic relationship between the system input and outputs, it may be used as the predictive model. However, in most cases it is difficult to obtain precise information about the system *a priori*. Moreover, even if it was available, the system may frequently vary its dynamics in its evolution over time. The purpose of adding an adaptive component to the predictive control system is to achieve, in an uncertain and time-varying environment, the results that would otherwise be obtained by the predictive control system only if the system dynamics were known.

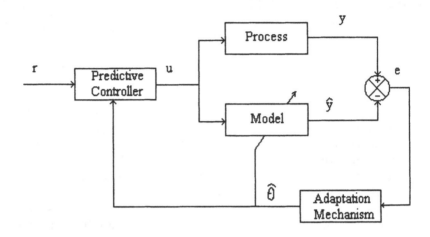

Fig. 7.1. Overall block diagram of an adaptive predictive control system

An adaptive predictive controller for a class of non-linear systems is obtained by combining parameter identification and predictive control algorithms, as shown in Fig. 7.1. The predictive model will change at each sampling instant t. In the adaptive system, the system model gives an estimation of the system output at instant t using model parameters also estimated at instant t. In the predictive controller, the estimated model is used to formulate the predictive model at instant t and to derive the control law.

Most predictive controllers presented in the literature, for example, GPC, extended horizon adaptive control (EHAC) and extended prediction self-adaptive control (EPSAC), are adaptive predictive controllers. The required model is obtained by using an identification algorithm, as shown in Fig. 7.1.

To estimate the parameters of a time-invariant linear system model from input/output data, the following descriptions will give a brief introduction to the parameter identification.

Consider a system to be exactly represented by a controlled autoregressive integrated moving average (CARIMA) model described by

$$a(q^{-1})y(t) = b(q^{-1})u(t - h) + \xi, \tag{7.1}$$

where $a(q^{-1}) = 1 + a_1 q^{-1} + ... + a_n q^{-n}$, and $b(q^{-1}) = b_1 q^{-1} + b_2 q^{-2} + ... + b_n q^{-n}$, and ξ denotes noise with zero mean. As stated in Chapter 4, the CARIMA model can be accommodated by a state space model which is used as a predictive model. Thus, the estimator is based on the CARIMA model. (7.1) can be rewritten as

$$y(t) = \varphi_0(t-1)^T \theta_0, \tag{7.2}$$

where

$$\varphi_0(t-1) = [y(t-1), ..., y(t-n), u(t-h-1), ..., u(t-h-n)]^T,$$
$$\theta_0 = [a_1, a_2, ..., a_n, b_1, b_2, ..., b_n]^T.$$

Now, at time $t+1$, with an additional set of data available, the objective is to determine the updated parameters $\hat{\theta}(t+1)$. Since the predictive model in the adaptive predictive control is updated at each sampling instant, $\hat{\theta}(t+1)$ should be identified recursively. A typical recursive identification scheme is the well-known recursive least squares algorithm, $i.e.$,

$$\hat{\theta}(t+1) = \hat{\theta}(t) + \frac{P(t)\varphi_0(t)}{1 + \varphi_0^T(t)P(t)\varphi_0}[y(t+1) - \varphi_0^T(t)\theta_0(t)], \tag{7.3}$$

$$P(t+1) = P(t) - \frac{P(t)\varphi_0(t)\varphi_0^T(t)P(t)}{1 + \varphi_0^T P(t)\varphi_0(t)}. \tag{7.4}$$

The updated parameters are used at each sampling instant by the predictive model. The resulting adaptive controller is sometimes called an indirect adaptive controller in contrast to a direct adaptive controller, where an estimator estimates the controller parameters directly (without first identifying a system model). Although good results have been reported using this method, analysis (for example, convergence analysis) of the resulting non-linear system as a whole in most reports is usually incomplete. Most often, the interaction between controller design and parameter identification is neglected. To solve this problem, Clarke et $al.$ (1994) discussed the stability of the adaptive predictive controller based on a linear time-invariant model. In this section, an adaptive predictive controller is developed and the stability of the closed-loop system is discussed giving consideration to the interconnection between predictive control and parameter estimates.

7.1.1 System Model

In the past two decades, significant progress has been made in the research and design of adaptive control systems. Fairly complete and comprehensive guidelines are now available for both design and implementation of adaptive controllers in cases where the system under control can be adequately modelled as a linear dynamical system. More recent developments in this area have been mainly focused and concerned with the design of adaptive controllers for non-linear dynamical systems.

It is generally acknowledged that adaptive control of the general single-input/single-output (SISO) non-linear system described by

$$y(t+1) = f(y(t), ..., y(t - n_y), u(t), ..., u(t - n_u)), \tag{7.5}$$

is a difficult and yet important problem to be addressed. There have been previous efforts to develop controllers for this class of non-linear systems. In Jordan and Rumelhart (1992), an inverse operator is necessary and invoked in the control design. Unfortunately, to date, the notion of an inverse operator for a non-linear system has not been well defined. If the control system is affine in the control effort (linear in u), or may be appropriately simplified to the form

$$y(t+1) = f(y(t), ..., y(t - n_y)) + g(y(t), ..., y(t - n_y))u(t), \tag{7.6}$$

non-linear control strategies such as that presented in Chen and Khalil (1995) are applicable.

In recent years, controllers in the form of a multilayer feedforward neural network have been developed (Goh, 1994) for general non-affine system (with unknown order and relative degree) in the state space formulation. However, these designs require as prior knowledge the bounds on the weights of the neural networks, but these requirements have remained a difficult practical issue to resolve. In Hou and Huang (1997), a model-free learning adaptive controller is presented using only the input and output information of the system. It is based on a new concept of partial derivative, called "pseudo-partial-derivative" (PPD) which may be derived from input and output measurements.

Consider the system described by the following SISO non-linear discrete-time equation

$$y(t+1) = f(y(t), ..., y(t - n_y), u(t), ..., u(t - n_u)), \tag{7.7}$$

where $y(t), u(t)$ are the output and input at time t, n_y, n_u are the unknown orders, and $f(.)$ is an unknown non-linear function.

Two assumptions are made first with regards to the system.

Assumption 7.1

The partial derivative of $f(\cdot)$ with respect to control input $u(t)$ is continuous.

Assumption 7.2

The sytem (7.7) is generalised Lipschitz, that is, satisfying $|\Delta y(t+1)| \leq C\|\Delta U(t)\|$ for $\forall t$ and $\|U(t)\| \neq 0$ where $\Delta y(t+1) = y(t+1) - y(t)$, $\Delta u(t) = u(t) - u(t-1)$ and $\Delta U(t) = [\Delta u(t), ..., \Delta u(t-L+1)]^T$, C is a constant, L is a positive integer known as the control input length constant of linearisation.

These assumptions for the system are reasonable and acceptable from a practical viewpoint. Assumption 7.1 is a typical condition for many control laws which a general non-linear system should satisfy. Assumption 7.2 poses a limitation on the rate of change of the system output permissible before the control law to be formulated is applicable. Theorem 7.1 will illustrate that this class of systems may be represented in an alternative and affine form.

Theorem 7.1

For the non-linear system (7.7), under assumptions (A1) and (A2), for a given L, there must exist $\Phi(t)$ and $\|\Phi(t)\| \leq C$ such that if $\|\Delta U(t)\| \neq 0$, the system may be described as :

$$\Delta y(t+1) = \Phi^T(t)\Delta U(t), \tag{7.8}$$

where $\Phi(t) = [\phi_1(t), ..., \phi_L(t)]^T$.

Proof

From (7.7),

$$\begin{aligned}
\Delta y(t+1) =\ & f(y(t), ..., y(t-n_y), u(t), ..., u(t-n_u)) \\
& -f(y(t-1), ..., y(t-n_y-1), u(t-1), ..., u(t-n_u-1)) \\
=\ & f(y(t), ..., y(t-n_y), u(t), ..., u(t-n_u)) \\
& -f(y(t), ..., y(t-n_y), u(t-1), ..., u(t-n_u-1)) \\
& +f(y(t), ..., y(t-n_y), u(t-1), ..., u(t-n_u-1)) \\
& -f(y(t), ..., y(t-n_y), u(t-2), ..., u(t-n_u-2)) \\
& \quad\ldots\ldots \\
& +f(y(t), ..., y(t-n_y), u(t-L+1), ..., u(t-n_u-L+1)) \\
& -f(y(t), ..., y(t-n_y), u(t-L), ..., u(t-n_u-L))
\end{aligned}$$

$$+f(y(t), ..., y(t-n_y), u(t-L), ..., u(t-n_u-L))$$
$$-f(y(t-1), ..., y(t-n_y-1), u(t-1), ..., u(t-n_u-1)).$$
$$(7.9)$$

By virtue of Assumption 7.1 and the Mean Value Theorem (Kroyszig, 1993), it follows that

$$\Delta y(t+1) = \frac{\partial f_*^T}{\partial u(t-1)} \Delta u(t) + \frac{\partial f_*^T}{\partial u(t-2)} \Delta u(t-1) + ...$$
$$+ \frac{\partial f_*^T}{\partial u(t-L)} \Delta u(t-L+1) + \psi(t), \qquad (7.10)$$

where $\frac{\partial f_*^T}{\partial u(t-i-1)} \Delta u(t-i)$ is the derivative of $f(...)$ with respect to u between $u(t-i-1)$ and $u(t-i)$, and

$$\psi(t) = \quad f(y(t), ..., y(t-n_y), u(t-L), ..., u(t-n_u-L))$$
$$-f(y(t-1), ..., y(t-n_y-1), u(t-1), ..., u(t-n_u-1)).$$

Since $\|U(t)\| \neq 0$, this implies that there exists at least one control $\Delta u(t-j) \neq 0$ $(0 \leq j \leq L-1)$. Consider the following equation:

$$\psi(t) = \eta^T(t)\Delta u(t-j). \qquad (7.11)$$

Since $\Delta u(t-j) \neq 0$, the solution to (7.11) always exists.

Denoting

$$\Phi^T(t) = [\phi_1(t), ..., \phi_{j-1}(t), ..., \phi_L(t)] = [\frac{\partial f_*^T}{\partial u(t-1)}, ..., \frac{\partial f_*^T}{\partial u(t-j-1)}$$
$$+\eta^T(t), ..., \frac{\partial f_*^T}{\partial u(t-L)}], \qquad (7.12)$$

(7.10) may be rewritten as

$$\Delta y(t+1) = \Phi^T(t)\Delta U(t). \qquad (7.13)$$

The theorem is proven.

The equation (7.8) may be rewritten as

$$x(t+1) = Ax(t) + B\Delta u(t), \qquad (7.14)$$
$$\Delta y(t) = c^T x(t), \qquad (7.15)$$

where

$$x(t) = [\Delta u(t-1)\ \Delta u(t-2)\ ...\Delta u(t-L)]^T, \tag{7.16}$$

$$A = \begin{bmatrix} 0\ 0\ ...\ 0\ 0 \\ 1\ 0\ ...\ 0\ 0 \\ 0\ 1\ ...\ 0\ 0 \\ \vdots\ \vdots\ ...\ 0\ 0 \\ 0\ 0\ ...\ 1\ 0 \end{bmatrix}, \tag{7.17}$$

$$B = [1\ 0\ 0...0\ 0]^T, \tag{7.18}$$

$$c = [\phi_1 \phi_2...\phi_{L-1}\ \phi_L]^T, \tag{7.19}$$

where A is an $L \times L$ matrix, B is an $L \times 1$ matrix, and c is an $L \times 1$ matrix. The parameter vector c can be identified on-line by using the leakage recursive least squares (RLS) algorithm (Haykin, 1996).

Remark 7.1

Theorem 7.1 is an extension of the results in Hou and Huang (1997). $\Phi(t)$ is a time-varying parameter vector even if the system (7.7) is a time-invariant system. This theorem shows that $\Phi(t)$ is a differential signal in some sense and bounded for any t. Thus, $\Phi(t)$ is slowly time-varying and its relation with $u(t)$ may be ignored when $\|\Delta U(t)\|$ and the sampling period are not too large.

7.1.2 Parameter Estimation

Based on the certainty equivalence principle, consider the design of a parameter identification scheme to identify the system I/O operator. For this purpose, consider the model (7.8) and Assumptions 7.1–7.2 and suppose that its order L is known. Now the system has the form:

$$\Delta y(t) = \phi_1(t)\Delta u(t-1) + \phi_2(t)\Delta u(t-2) + ... + \phi_L(t)\Delta u(t-L),$$

which, in turn, is parametrised in terms of a vector c as

$$c = [\phi_1 \phi_2...\phi_L]^T. \tag{7.20}$$

The parameters are related to the coefficients of $\Delta u(t)$. Thus, the system is estimated by the parametric model:

$$\Delta \hat{y} = \hat{c}^T \Delta U(t-1). \tag{7.21}$$

Employing the previous results, the system can be identified by estimating the unknown parameters with a suitable adaptive law. For example, if c is a constant, the following standard recursive least squares algorithm can be used:

$$\hat{c}(t) = \hat{c}(t-1) + \frac{\nu \Delta U(t-1)}{\mu + \Delta U^T(t-1)\Delta U(t-1)}[\Delta y(t)$$
$$-\Delta U^T(t-1)\hat{c}(t-1)],$$

where $\nu > 0$ and $\mu > 0$. However, due to the time-varying nature, a modified adaptive law is more suitable for the present system:

$$\hat{c}(t) = \varrho\hat{c}(t-1) + \frac{\nu \Delta U(t-1)}{\mu_1 + \mu_2 \Delta U^T(t-1)\Delta U(t-1)}[\Delta y(t)$$
$$-\Delta U^T(t-1)\hat{c}(t-1)], \qquad (7.22)$$

where ϱ is very close to $1, \nu > 0, \mu_1 > 0$ and $\mu_2 > max\{\nu/\varrho, \frac{\nu L}{\varpi-\varrho}\}$ with a constant ϖ subject to $1 > \varpi > \varrho$.

Having obtained the parameter estimates , the identified model can be used to formulate the predictive model at instant t. This is discussed in the next subsection.

7.1.3 Prediction and Control Action

The structure of the model (7.14),(7.15) is useful in the formulation of predictive controllers (see Clarke *et al.*, 1987). First define a state prediction model of the form:

$$\hat{x}(t+j/t) = A\hat{x}(t+j-1/t) + B\Delta u(t+j-1/t), \quad j = 1, 2, ..., p+1, (7.23)$$

where $\hat{x}(t + j/t)$ denotes the state vector prediction at instant t for instant $t + j$ and $u(\cdot/t)$ denotes the sequence of control vectors on the prediction interval. This model is redefined at each sampling instant t from the actual state vector and the controls previously applied,

$$\hat{x}(t/t) = x(t); \quad u(t - j/t) = u(t - j); \quad j = 1, 2, ..., h. \qquad (7.24)$$

Consider a linear quadratic performance index written in the following form:

$$J = \frac{1}{2}\sum_{j=1}^{p+1}[y_r(t+j) - \hat{y}(t+j)]^T Q_j[y_r(t+j) - \hat{y}(t+j)]$$

$$+\frac{1}{2}\sum_{j=0}^{p}\Delta u(t+j)^{T}R_{j}\Delta u(t+j), \qquad (7.25)$$

where $y_r(t+j)$ is a reference trajectory for the output vector which may be redefined at each sampling instant t, $\hat{y}(t+j)$ is a prediction output trajectory and $\Delta u(t+j) = u(t+j) - u(t+j-1)$. Q, R are symmetric weighting matrices.

Applying (7.23) recursively to the initial conditions, it follows that

$$\left. \begin{aligned} \hat{y}(t+1/t) - \hat{y}(t/t) &= c^{T}Ax(t) + c^{T}B\Delta u(t), \\ \hat{y}(t+2/t) - \hat{y}(t+1/t) &= c^{T}A^{2}x(t) + c^{T}AB\Delta u(t) \\ &\quad + c^{T}B\Delta u(t+1), \\ &\vdots \\ \hat{y}(t+p+1/t) - \hat{y}(t+p/t) &= c^{T}A^{p+1}x(t) + c^{T}A^{p}B\Delta u(t) + ... \\ &\quad + c^{T}B\Delta u(t+p). \end{aligned} \right\} \quad (7.26)$$

Summing both sides of (7.26) yields:

$$\hat{y}(t+p+1) = y(t) + GAx(t) + FU, \qquad (7.27)$$

where

$$G = \left[c^{T} + c^{T}A + ... + c^{T}A^{p+1} \right], \qquad (7.28)$$
$$F = [(c^{T}B + c^{T}AB + ... + c^{T}A^{p}B)\, (c^{T}B + c^{T}AB + ... + c^{T}A^{p-1}B)...$$
$$c^{T}B], \qquad (7.29)$$
$$U = [\Delta u(t)\ \Delta u(t+1)\ ...\Delta u(t+p)]^{T}. \qquad (7.30)$$

The main difficulty associated with a predictive control design based on (7.26) is due to the number of unknowns p, corresponding to the number of values in the control sequence $\Delta u(t+j)$. One way of reducing the number of unknowns is to predetermine the form of the control sequence. To this end, the control sequence is made constant over the prediction interval:

$$\Delta u(t+1) = \Delta u(t+2) = ... = \Delta u(t+p) = 0. \qquad (7.31)$$

Thus, the prediction equation with a $t+p+1$ horizon is given by

$$\hat{y}(t+p+1) = y(t) + GAx(t) + \bar{F}\Delta u(t), \qquad (7.32)$$

where

$$\bar{F} = [c^T B + c^T A B + c^T A^2 B + ... + c^T A^{p-1}]. \tag{7.33}$$

Finally, consider the following selection of weighting vectors in the performance index (7.25):

$$Q_j = 0, Q_{t+p+1} = 1, \tag{7.34}$$
$$R_j = 0, R_0 = r, \tag{7.35}$$

for $j = 1, ..., p + 1$. Given this selection, the performance index reduces to

$$J = \frac{1}{2}[y_r(t + p + 1) - \hat{y}(t + p + 1)]^2 + \frac{1}{2}r\Delta u(t)^2. \tag{7.36}$$

Predictive control is capable of coping with both constant and varying future set-points. Substituting (7.32) into (7.36), the solution minimising the performance index may then be obtained by solving

$$\frac{\partial J}{\partial \Delta u(t)} = 0, \tag{7.37}$$

from which direct computations may be applied to yield the control law explicitly as

$$\Delta u(t) = (\bar{F}^T \bar{F} + r)^{-1} \bar{F}^T [y_r(t + p + 1) - y(t) - GAx(t)]. \tag{7.38}$$

Remark 7.2

The main feature of the result obtained is an adaptive predictive control based on the conventional GPC concepts. This differs from Lee and Yu (1997) and Zheng and Morari (1993) whose results are non-adaptive versions. In addition, although the result in Lee and Yu (1997) is designed for a general time-varying system, the stability can be guaranteed only for the infinite-horizon case. This cannot be implemented directly as the cost is unbounded due to the presence of the integrator (see Section 3.2 of Lee and Yu, 1997). To solve this problem, Lee and Yu (1997) derived a sub-optimal algorithm under the constrained receding horizon control (CRHC) for a particular class of finite impulse response (FIR) systems. Recently, Zelinka et al.(1999) showed that the CRHC is more sensitive to system time delay and uncertainties in comparison with the conventional GPC. The same comments apply for Mayne and Michalska (1990), Mayne and Michalska (1993), and Zheng and Morari (1993), since their result is also a CRHC control.

7.2 Stability Analysis

For the proposed adaptive controller, it is interesting to examine the closed-loop stability properties of the system. In the following development, stability results pertaining to the adaptive predictive control system are provided. Before giving the stability result, parameter convergence is discussed first.

7.2.1 Parameter Identification Convergence

In the following the adaptation mechanism defined in (7.22) is proved to be capable of guaranteeing the parameter boundedness of the identification algorithm. This property will be useful for the stability anlysis of the closed-loop system.

Theorem 7.2

For the system (7.8), if the modified RLS algorithm (7.22) is used to identify the model parameter vector c, the estimated parameter $\hat{c}$ is bounded.

Proof

To prove that the estimated $\hat{c}(t)$ in (7.22) is bounded, the following notation is introduced:

$$w(t) = c(t) - c(t-1), \tag{7.39}$$
$$\tilde{c}(t) = \hat{c}(t) - c(t). \tag{7.40}$$

Since $\|\Phi(t)\| \leq C$ (Theorem 7.1), it follows that

$$\|w(t)\| = \|c(t) - c(t-1)\| = \|\Phi(t-1) - \Phi(t-2)\| \leq 2C, \tag{7.41}$$

i.e., $w(t)$ is bounded.

From (7.22), it is known that

$$
\begin{aligned}
\tilde{c}(t) &= \hat{c}(t) - c(t) = \varrho\hat{c}(t-1) - \varrho c(t-1) - (1-\varrho)c(t-1) - w(t) \\
&\quad + \frac{\nu\Delta U(t-1)}{\mu_1 + \mu_2\Delta U^T(t-1)\Delta U(t-1)}[\Delta U^T(t-1)c(t-1) \\
&\quad - \Delta U^T(t-1)\hat{c}(t-1)] \\
&= \varrho\tilde{c}(t-1) - \frac{\nu\Delta U(t-1)}{\mu_1 + \mu_2\Delta U^T(t-1)\Delta U(t-1)}\Delta U^T(t-1)\tilde{c}(t-1)
\end{aligned}
$$

$$-(1-\varrho)c(t-1)-w(t)$$
$$= [\varrho I - \frac{\nu \Delta U(t-1)}{\mu_1 + \mu_2 \Delta U^T(t-1)\Delta U(t-1)} \Delta U^T(t-1)]\tilde{c}(t-1)$$
$$-(1-\varrho)c(t-1)-w(t). \tag{7.42}$$

Let

$$\Xi(t) = \varrho I - \frac{\nu \Delta U(t-1)}{\mu_1 + \mu_2 \Delta U^T(t-1)\Delta U(t-1)} \Delta U^T(t-1).$$

To show the boundeness of $\hat{c}(t)$, it is necessary to prove that $||\Xi(t)|| < 1$. Computing $||\Xi(t)||$ yields

$$||\Xi(t)|| = \max_k [|\varrho - \frac{\nu \Delta u^2(t-k)}{\mu_1 + \mu_2 ||\Delta U(t-1)||^2}|$$
$$+ \frac{\nu |\Delta u(t-k)| \sum_{i=1}^{L} |\Delta u(t-i)|}{\mu_1 + \mu_2 ||\Delta U(t-1)||^2}].$$

Since

$$\frac{\nu \Delta u^2(t-k)}{\mu_1 + \mu_2 \Delta U^T(t-1)\Delta U(t-1)} \le \frac{\nu \Delta u^2(t-k)}{\mu_1 + \mu_2 \Delta u^2(t-k)},$$

and $\mu_1 > 0, \varrho\mu_2 > \nu > 0$, it follows that

$$0 < \frac{\nu \Delta u^2(t-k)}{\mu_1 + \mu_2 ||\Delta U(t-1)||^2} < \varrho. \tag{7.43}$$

This implies that

$$||\Xi(t)|| = \max_k [\varrho - \frac{\nu \Delta u^2(t-k)}{\mu_1 + \mu_2 ||\Delta U(t-1)||^2}$$
$$+ \frac{\nu |\Delta u(t-k)| \sum_{i=1}^{L} |\Delta u(t-i)|}{\mu_1 + \mu_2 ||\Delta U(t-1)||^2}]$$

$$= \max_k [\varrho + \frac{-\nu \Delta u^2(t-k) + \nu |\Delta u(t-k)| \sum_{i=1}^{L} |\Delta u(t-i)|}{\mu_1 + \mu_2 ||\Delta U(t-1)||^2}].$$

ϖ is a constant where $1 > \varpi > \varrho$ (note that ϖ can always be found). Now the problem is to prove that

$$\frac{-\nu\Delta u^2(t-k) + \nu|\Delta u(t-k)|\sum_{i=1}^{L}|\Delta u(t-i)|}{\mu_1 + \mu_2||\Delta U(t-1)||^2} < \varpi - \varrho. \qquad (7.44)$$

Let $u_{max} = max_{1\leq j\leq L}[|\Delta u(t-j)|]$. It follows that

$$\nu|\Delta u(t-k)|\sum_{i=1}^{L}|\Delta u(t-i)| < \nu L u_{max}^2, \qquad (7.45)$$

$$||\Delta U(t-1)||^2 = \sum_{j=1}^{L}\Delta u^2(t-j) > u_{max}^2. \qquad (7.46)$$

Since $\mu_2 > max\{\nu/\varrho, \frac{\nu L}{\varpi-\varrho}\}$, then $(\varpi - \varrho)\mu_2 > \nu L$. This implies that

$$\nu\Delta u^2(t-k) + (\varpi - \varrho)\mu_1 + (\varpi - \varrho)\mu_2||U(t-1)||^2$$
$$> \nu L u_{max}^2 > \nu|\Delta u(t-k)|\sum_{i=1}^{L}|\Delta u(t-i)|.$$

This implies that (7.44) holds. This also implies that

$$||\Xi(t)|| = \max_{k}[\varrho - \frac{\nu\Delta u^2(t-k)}{\mu_1 + \mu_2||\Delta U(t-1)||^2}$$
$$+ \frac{\nu|\Delta u(t-k)|\sum_{i=1}^{L}|\Delta u(t-i)|}{\mu_1 + \mu_2||\Delta U(t-1)||^2}] < \varpi < 1.$$

Taking the norm of both sides of (7.42) and using Assumption 7.2 yields

$$||\tilde{c}(t)|| \leq ||\Xi(t)||\,||\,\tilde{c}(t-1)\,|| + (3-\varrho)C, \qquad (7.47)$$

$$||\tilde{c}(t)|| < \varpi^t\,||\,\tilde{c}(0)\,|| + \frac{1-\varpi^t}{1-\varpi}(3-\varrho)C, \qquad (7.48)$$

$$\lim_{t\to\infty}||\tilde{c}(t)|| < \frac{(3-\varrho)C}{1-\varpi}. \qquad (7.49)$$

Since $||c(t)|| \leq C$ is bounded, $\hat{c}(t)$ is also bounded. This completes the proof.

7.2.2 Stability of Adaptive Controller

Consider the actual perturbed system described by

$$z(t+1) = (A + \delta(t)A_1(t))z(t) + \delta(t)A_2(t)w(t), \qquad (7.50)$$

where $w(t)$ is a perturbed vector, $\delta(t) < \delta_0$ (δ_0 is a positive constant) and $||A_1(t)|| < M_1, ||A_2(t)|| < M_2$.

Define a Lyapunov function

$$V(z(t)) = z^T(t)Pz(t), \tag{7.51}$$

where P is the solution of

$$A^T PA - P = -I, \tag{7.52}$$

and I is the unit matrix.

Lemma 7.1

For the system described by (7.50), if δ_0 is sufficiently small such that

$$\delta_0 \le \min\{\frac{1}{[2||A^T||M_1 + M_1^2 + (||A|| + M_1)M_1||P||]||P||},$$
$$\frac{1 - \rho\lambda_{max}(P)}{[2||A^T||M_1 + M_1^2 + (||A|| + M_1)M_1||P||]||P||}, 1\}, \tag{7.53}$$

and $0 < \rho < 1$ is a constant, then

$$\sqrt{V(t+1)} < \sqrt{(1-\rho)}\sqrt{V(t)} + \sqrt{\delta_0 h_0}||w(t)||, \tag{7.54}$$

where

$$h_0 = M_2^2 + M_2^2||P||. \tag{7.55}$$

Proof

According to the Lyapunov function,

$$\lambda_{min}(P)||z(t)||^2 \le V(z(t)) \le \lambda_{max}(P)||z(t)||^2. \tag{7.56}$$

It can be directly shown that

$$\Delta V(z(t)) = z^T(t+1)Pz(t+1) - z(t)^T Pz(t)$$
$$= z^T(t)[(A + \delta(t)A_1(t))^T P(A + \delta(t)A_1(t)) - P]z(t)$$
$$+ 2\delta(t)z^T(t)(A + \delta(t)A_1(t))^T PA_2(t)w(t)$$

$$+\delta^2(t)w^T(t)A_2^T(t)PA_2(t)w(t)$$
$$= z^T(t)[A^TPA - P + 2\delta(t)A^TPA_1(t)$$
$$+\delta^2(t)A_1^T(t)PA_1(t)]z(t)$$
$$+2\delta(t)z^T(t)(A + \delta A_1(t))^TPA_2(t)w(t)$$
$$+\delta(t)^2 w^T(t)A_2^T(t)PA_2^T(t)w(t)$$
$$= z^T(t)[-I + 2\delta(t)A^TPA_1(t) + \delta(t)^2 A_1^T(t)PA_1(t)]z(t)$$
$$+2\delta(t)z^T(t)(A + \delta A_1(t))^TPA_2(t)w(t)$$
$$+\delta(t)^2 w^T(t)A_2^T(t)PA_2^T(t)w(t). \tag{7.57}$$

By using the following inequality for the vectors ξ, and ζ,

$$2\xi^T\zeta \le \xi^T\xi + \zeta^T\zeta, \tag{7.58}$$

it follows that

$$2\delta(t)z^T(t)(A + \delta A_1(t))^TPA_2(t)w(t) \le \delta(t)z^T(t)(A + \delta A_1(t))^T$$
$$\times PPA_1(t)z(t)$$
$$+\delta(t)w^T(t)A_2^T(t)A_2(t)w(t).$$

Substituting the above equation into (7.57) yields:

$$\Delta V(t) \le z^T(t)[-I + 2\delta(t)A^TPA_1(t) + \delta(t)^2 A_1^T(t)PA_1(t)$$
$$+\delta(t)(A + \delta A_1(t))^TPPA_1(t)]z(t)$$
$$+\delta(t)w^T(t)[A_2^T(t)A_2(t) + \delta A_2^T(t)PA_2(t)]w(t)$$
$$\le -||z(t)||^2 + \delta(t)z^T(t)[2\delta(t)A^TPA_1(t) + \delta(t)A_1^T(t)PA_1(t)$$
$$+(A + \delta A_1(t))^TPPA_1(t)]z(t)$$
$$+\delta(t)w^T(t)[A_2^T(t)A_2(t) + \delta A_2^T(t)PA_2(t)]w(t).$$

Next, it is noted that

$$\delta(t)z^T(t) \ [\ 2A^TPA_1(t) + \delta(t)A_1^T(t)PA_1(t)$$
$$+(A + \delta A_1(t))^TPPA_1(t)]z(t)$$
$$\le \delta_0[2||A^TP||||A_1(t)|| + \delta_0||A_1(t)||^2||P|| + (||A||$$
$$+\delta_0||A_1(t)||)||A_1(t)||||P||]||z(t)||^2$$
$$\le \delta_0[2||A^T||M_1 + \delta_0 M_1^2 + (||A|| + \delta_0 M_1)M_1||P||]||P||||z(t)||^2.$$

Since δ_0 is sufficiently small, it follows that $\delta_0 < 1$. Thus, it may be inferred that

$$\delta(t)z^T(t) \ [\ 2A^T PA_1(t) + \delta(t)A_1^T(t)PA_1(t)$$
$$+(A + \delta A_1(t))^T PPA_1(t)]z(t)$$
$$\leq \delta_0[2||A^T||M_1 + M_1^2 + (||A|| + M_1)$$
$$\times M_1||P||]||P||||z(t)||^2. \tag{7.59}$$

Similarly,

$$\delta(t)w^T(t)[A_2^T(t)A_2(t) + \delta A_2^T(t)PA_2(t)]w(t)$$
$$\leq \delta_0(M_2^2 + M_2^2||P||)||w(t)||^2. \tag{7.60}$$

Denoting

$$\delta_1 = 1 - \delta_0[2||A^T||M_1 + M_1^2 + (||A|| + M_1)M_1||P||]||P||, \tag{7.61}$$

it follows from the inequality (7.56) that

$$\Delta V(t) \leq -\delta_1||z(t)||^2 + \delta_0 h_0||w(t)||^2$$
$$\leq -\frac{\delta_1}{\lambda_{max}(P)}V(t) + \delta_0 h_0||w(t)||^2.$$

Since the condition (7.53) holds, it follows that

$$V(t+1) < (1 - \rho)V(t) + \delta_0 h_0||w(t)||^2$$
$$\leq (1 - \rho)V(t) + 2\sqrt{(1 - \rho)V(t)\delta_0 h_0||w(t)||} + \delta_0 h_0||w(t)||^2$$
$$= (\sqrt{(1 - \rho)V(t)} + \sqrt{\delta_0 h_0}||w(t)||)^2. \tag{7.62}$$

Since $V(t) \geq 0$ for $t \geq 0, 1 - \rho > 0$, and $\delta_0, h_0 > 0$.

Lemma 7.2 (Elshafei et al., 1995)

Let

$$z(t+1) = Az(t), \tag{7.63}$$

such that

$$A^T PA - P = -Q. \tag{7.64}$$

Then, the perturbed system

$$z(t+1) = (A + \epsilon)z(t), \tag{7.65}$$

is asymptotically stable if

$$0 \leq ||\epsilon|| < -||A|| + \sqrt{||A||^2 + \frac{\lambda_{min}(Q)}{||P||}}. \tag{7.66}$$

Lemma 7.3 (Payne, 1987)

Consider the time-varying difference equation

$$z(t+1) = \Gamma(t)z(t) + v(t), \quad t \geq 0, \tag{7.67}$$

where $z(t)$ and $v(t)$ are real vectors of finite dimension. Suppose that the sequence of matrices $\{\Gamma(t)\}$ and $z(0) = z_0$ are bounded and that the free system

$$z(t+1) = \Gamma(t)z(t), \quad t \geq 0, \tag{7.68}$$

is exponentially stable. Furthermore, suppose that there exist sequences of non-negative real numbers $\{\beta(t)\}$ and $\{\sigma(t)\}$ and an integer $N \geq 0$ such that

$$||v(t+j)|| \leq \beta(t) \sum_{i=0}^{N} ||z(t+j-i)|| + \sigma(t+j), \tag{7.69}$$

where $|| \bullet ||$ denotes the usual Euclidean norm. Under these conditions, if $\{\beta(t)\}$ converge to zero and $\{\sigma(t)\}$ is bounded, then $\{z(t)\}$ and $\{v(t)\}$ are bounded.

Theorem 7.3

Under Assumptions 7.1 and 7.2, suppose that the system described by (7.14) and (7.15) is controlled by (7.38) and the estimate $\hat{c}$ is identified using the leakage recursive least squares algorithm (7.22). If $y_{sp} = Const, y_r(t+p+1) = \alpha^t y_{sp} + (1 - \alpha^t)y(t)$ $(0.9 < \alpha < 1)$ and r is large, then

(1) the asymptotic output tracking error $y(t) - y_{sp}$ is bounded;

(2) $\{y(t)\}, \{u(t)\}$ are bounded sequences.

Proof

Since the boundedness of $\hat{c}(t)$ has been proved in (7.22), the result is available to prove that the tracking error and system output and input are bounded.

Based on the estimated value of $c(t)$, first prove that the tracking error is bounded. Let

$$e(t) = y_{sp} - y(t). \tag{7.70}$$

Thus, the control law (7.38) may be written as

$$\Delta u(t) = (\bar{F}^T \bar{F} + r)^{-1} \bar{F}^T [\alpha^t e(t) - GAx(t)]. \tag{7.71}$$

Let

$$\beta = (\bar{F}^T \bar{F} + r)^{-1}. \tag{7.72}$$

Applying (7.71) to (7.14) yields

$$x(t+1) = (A - \beta \bar{F}^T GA)x(t) + \beta F^T \alpha^t e(t). \tag{7.73}$$

Since $F^T GA$ is bounded and $\beta \le \beta_0 = r^{-1}$, when r is large, using Lemma 7.1, gives

$$\sqrt{V(2)} < a\sqrt{V(1)} + \beta_0 H\alpha^1 e(1), \tag{7.74}$$

$$\begin{aligned}\sqrt{V(3)} < {} & a^2\sqrt{V(1)} + a\beta_0 H\alpha^1 e(1) \\ & + \beta_0 H\alpha^2 e(2),\end{aligned} \tag{7.75}$$

$$\begin{aligned}\sqrt{V(4)} < {} & a^3\sqrt{V(1)} + a^2\beta_0 H\alpha^1 e(1) + a\beta_0 H\alpha^2 e(2) \\ & + \beta_0 H\alpha^3 e(3),\end{aligned} \tag{7.76}$$

......

$$\begin{aligned}\sqrt{V(t)} < {} & a^{t-1}\sqrt{V(1)} + a^{t-2}\beta_0 H\alpha^1 e(1) + ... + a\beta_0 H\alpha^{t-2} e(t-2) \\ & + \beta a_0 H\alpha^{t-1} e(t-1),\end{aligned} \tag{7.77}$$

where $0 < a < 1$ and H is constant. Summing the two sides of the above equations, results in

$$\begin{aligned}\sqrt{V(2)} + \sqrt{V(3)} + ... + \sqrt{V(t)} < {} & \frac{a - a^t}{1 - a}\sqrt{V(1)} \\ & + \frac{1 - a^{t-1}}{1 - a}\beta_0 H\alpha^1 e(1) \\ & + \frac{1 - a^{t-2}}{1 - a}\beta_0 H\alpha^2 e(2) + ... \\ & + \frac{1 - a^2}{1 - a}\beta_0 H\alpha^{t-2} e(t-2) \\ & + \frac{1 - a}{1 - a}\beta_0 H\alpha^{t-1} e(t-1).\end{aligned}$$

Since $0 < a < 1$, $-a^t < 0$, it follows that

$$
\sqrt{V(2)} + \sqrt{V(3)} + ... + \sqrt{V(t)} < \frac{a}{1-a}\sqrt{V(1)} + \frac{1}{1-a}\beta_0 H a^1 e(1)
$$
$$
+ \frac{1}{1-a}\beta_0 H a^2 e(2) + ... + \frac{1}{1-a}\beta_0 H a^{t-2} e(t-2)
$$
$$
+ \frac{1}{1-a}\beta_0 H a^{t-1} e(t-1)
$$
$$
\leq \frac{a}{1-a}\sqrt{V(1)} + \beta_0 \frac{\alpha - \alpha^t}{(1-\alpha)(1-a)} \max\{e(\tau)|1 \leq \tau \leq t-1\}
$$
$$
< \frac{a}{1-a}\sqrt{V(1)} + \beta_0 \frac{\alpha}{(1-\alpha)(1-a)} \max\{e(\tau)|1 \leq \tau \leq t-1\}
$$
$$
(\text{from} - \alpha^t < 0). \tag{7.78}
$$

From (7.56), it follows that

$$
\sqrt{\lambda_{min}(P)}||x(t)|| \leq \sqrt{V(t)}. \tag{7.79}
$$

Applying the above inequality to (7.78), it follows that

$$
||x(2)|| + ||x(3)|| + ... + ||x(t)|| < \frac{a}{(1-a)\sqrt{\lambda_{min}(P)}}\sqrt{V(1)}
$$
$$
+ \beta_0 \frac{\alpha}{\sqrt{\lambda_{min}(P)}(1-\alpha)(1-a)} \max\{e(\tau)|1 \leq \tau \leq t-1\}. \tag{7.80}
$$

Assumption 7.2 implies that

$$
|e(t)| - |e(t-1)| \leq |y(t) - y(t-1)| \leq C||x(t)||. \tag{7.81}
$$

Thus,

$$
|e(2)| \leq |e(1)| + C||x(2)||,
$$
$$
|e(3)| \leq |e(2)| + C||x(3)||,
$$
$$
...
$$
$$
|e(t)| \leq |e(t-1)| + C||x(t)||,
$$
$$
\Rightarrow
$$
$$
|e(t)| \leq |e(1)| + C(||x(2)|| + ||x(3)|| + ... + ||x(t)||). \tag{7.82}
$$

Substituting (7.80) into the above yields:

$$
|e(t)| \leq |e(1)| + C[\frac{a}{(1-a)\sqrt{\lambda_{min}(P)}}\sqrt{V(1)}
$$

$$+\beta_0 \frac{\alpha}{\sqrt{\lambda_{min}(P)}(1-\alpha)(1-a)} \max\{e(\tau)|1 \leq \tau \leq t-1\}]$$
$$\leq C_0 + \beta_0 C_1 \max\{e(\tau)|1 \leq \tau \leq t-1\},$$

where

$$C_0 = |e(1)| + C[\frac{a}{(1-a)\sqrt{\lambda_{min}(P)}}\sqrt{V(1)}], \tag{7.83}$$

$$C_1 = C\frac{\alpha}{\sqrt{\lambda_{min}(P)}(1-\alpha)(1-a)}. \tag{7.84}$$

Let

$$e_\tau = max\{e(\tau)|1 \leq \tau \leq t-1\},$$

$$|e(t+1)| \leq C_0 + \beta_0 C_1 max_{1 \leq i \leq t}\{e_\tau, e(t)\}$$
$$\begin{cases} \leq & C_0 + \beta_0 C_1 e_\tau \ \ if \ e_\tau \geq e(t) \\ < C_0 + \beta_0 C_1 C_0 + \beta_0^2 C_1^2 e_\tau \ \ if \ e_\tau < e(t) \end{cases}$$

$$|e(t+2) \ | \begin{cases} \leq & C_0 + \beta_0 C_1 max_{1 \leq i \leq t}\{e_\tau, e(t)\} \\ & \begin{cases} \leq C_0 + \beta_0 C_1 e_\tau \ \ if \ e_\tau \geq e(t) \\ < C_0 + \beta_0 C_1 C_0 + \beta_0^2 C_1^2 e_\tau \ \ if \ e_\tau < e(t) \end{cases} \\ < & C_0 + \beta_0 C_1 C_0 + \beta_0^2 C_1^2 max_{1 \leq i \leq t}\{e_\tau, e(t)\} < \\ & \begin{cases} C_0 + \beta_0 C_1 C_0 + \beta_0^2 C_1^2 e_\tau \ \ if \ e_\tau \geq e(t) \\ C_0 + \beta_0 C_1 C_0 + \beta_0^2 C_1^2 C_0 + \beta_0^3 C_1^3 e_\tau \ \ if \ e_\tau < e(t) \end{cases} \end{cases}$$

In summary,

$$|e(t+2)| \begin{cases} \leq & C_0 + \beta_0 C_1 e_\tau \\ < & C_0 + \beta_0 C_1 C_0 + \beta_0^2 C_1^2 e_\tau \\ < C_0 + \beta_0 C_1 C_0 + \beta_0^2 C_1^2 C_0 + \beta_0^3 C_1^3 e_\tau \end{cases} \tag{7.85}$$

Repeatedly using the procedure, after q times the following relation will hold:

$$|e(t+q)| \begin{cases} \leq & C_0 + \beta_0 C_1 e_\tau \\ < & C_0 + \beta_0 C_1 C_0 + \beta_0^2 C_1^2 e_\tau \\ < & C_0 + \beta_0 C_1 C_0 + \beta_0^2 C_1^2 C_0 + \beta_0^3 C_1^3 e_\tau \\ ... \\ < C_0[1 + \beta_0 C_1 + (\beta_0 C_1)^2 + (\beta_0 C_1)^3 + ... \\ \quad +(\beta_0 C_1)^q] + \beta_0^{q+1} C_1^{q+1} e_\tau \end{cases} \tag{7.86}$$

Since

$$C_0 < C_0[1 + \beta_0 C_1 + (\beta_0 C_1)^2 + (\beta_0 C_1)^3 + \ldots$$
$$+ (\beta_0 C_1)^q],$$
$$C_0 + \beta_0 C_1 C_0 < C_0[1 + \beta_0 C_1 + (\beta_0 C_1)^2 + (\beta_0 C_1)^3 + \ldots$$
$$+ (\beta_0 C_1)^q],$$
$$C_0 + \beta_0 C_1 C_0 + \beta_0^2 C_1^2 C_0 < C_0[1 + \beta_0 C_1 + (\beta_0 C_1)^2 + (\beta_0 C_1)^3 + \ldots$$
$$+ (\beta_0 C_1)^q],$$

$$\ldots$$

and

$$\beta_0^{q+1} C_1^{q+1} < \beta_0^q C_1^q < \ldots < \beta_0^2 C_1^2 < \beta_0 C_1,$$
(since β_0 is sufficiently small, $0 < \beta_0 C_1 < 1$)

it follows that

$$|e(t+q)| < C_0 \frac{1 - (\beta_0 C_1)^{q+1}}{1 - \beta_0 C_1} + \beta_0 C_1 e_\tau$$
$$< C_0 \frac{1}{1 - \beta_0 C_1} + \beta_0 C_1 e_\tau \quad (\text{since} - (\beta_0 C_1)^{q+1} < 0). \quad (7.87)$$

This implies that $e(t+q)$ is bounded when $q \to \infty$.

(2) Since y_{sp} is constant and $e(t)$ is bounded, $y(t)$ is also bounded. From Lemma 7.2, the free system of (7.73) is asymptotically stable since F, G are bounded and $\beta(t)$ is sufficiently small. Since $\beta F^T \alpha^t e(t)$ is bounded, $x(t)$ is bounded by virtue of Lemma 7.3. This implies $\Delta u(t)$ is bounded.

From the following inequality,

$$|u(t)| \leq |u(t) - u(1)| + |u(1)|$$
$$\leq |\Delta u(t)| + |\Delta u(t-1)| + \ldots + |\Delta u(2)| + |u(1)|, \quad (7.88)$$

it can be concluded that $u(t)$ is bounded.

Remark 7.3

The choice of μ_2 is $\mu_2 > max\{\nu/\varrho, \frac{\nu L}{\varpi - \varrho}\}$ in order to achieve (7.44). For a practical control problem, it is not necessary to satisfy this condition since the input control is always bounded. Typically, the selection of $\mu_2 = 1$ and a large μ_1 can be made to achieve (7.44) (this will be shown in the simulation section).

7.3 Simulation

In this section, three simulation examples on real practical control problems are provided to further illustrate the principles of the adaptive predictive controller and demonstrate its effectiveness.

7.3.1 Linear System

This example illustrates the applicability of the proposed control to linear systems. Consider a linear system described by

$$x(t+1) = \begin{bmatrix} 1.8822 & 1.000 \\ -0.8842 & 0 \end{bmatrix} x(t) + \begin{bmatrix} 0.0035 \\ 0.0034 \end{bmatrix} u(t). \tag{7.89}$$

The following equation is used to model (7.89):

$$\Delta y(t+1) = c_0(t)\Delta u(t) + c_1(t)\Delta u(t-1) + c_2(t)\Delta u(t-2)$$
$$+c_3\Delta u(t-3) + c_4(t)\Delta u(t-4).$$

Applying the controller (7.38) yields the control response shown in Fig. 7.2, where the leakage factor $\varrho = 0.99$, the control weighting r is chosen as 250, and the prediction horizon p is selected as $p = 10$.

7.3.2 The Injection Moulding Control Problem

The injection moulding problem given in Chapter 4 is used again in this chapter. This application example will be mainly concerned with the non-linear system control of the filling phase, while the preceding chapters focused on the linear system. The entire phase may be regulated by controlling the injection speed of the ram so as to follow a pre-generated trajectory.

The reader is referred to Wei *et al.* (1994) for the general theory of injection moulding, and to Rafizadeh *et al.* (1996) for the related control problems. A non-linear dynamic model of the injection moulding process governing the ram velocity to the hydraulic oil flow during the filling phase is described by

$$\dot{z} = v_z,$$
$$\dot{P_1} = \frac{\beta_1}{v_{10} + A_1 z}(u - A_1 v_z),$$

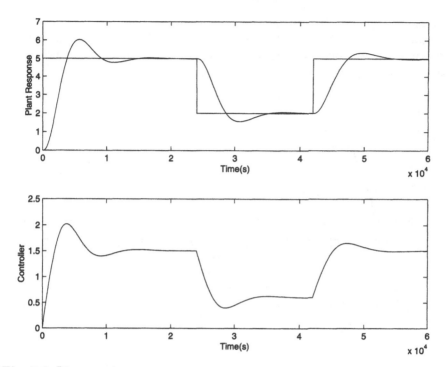

Fig. 7.2. Linear system response

$$\dot{v}_z = \frac{1}{M}[P_1A_1 - P_2A_2 - 2\pi\eta R_n^{1-n}(l_0 + z)(\frac{(s-1)v_z}{k_r^{1-s} - 1})^n],$$

$$\dot{P}_2 = \frac{\beta_2}{v_{20} - A_2 z}(A_2 v_z - Q_p),$$

$$y = z,$$

where z is the ram displacement, v_z is the ram velocity, P_1 is the hydraulic pressure, P_2 is the nozzle pressure, Q_p is the polymer flow rate, and u is the hydraulic oil flow into the injection cylinder. $\beta_1, \beta_2, v_{10}, v_{20}, A_1, A_2, M, \eta, R_n,$ $l_0, s, k_r,$ and n are system parameters. The physical interpretation of these parameters and their values are shown in Tan (1999). In this example, it will be assumed that these parameters are unknown and to be determined. In addition, it is assumed that during the filling phase, the polymer melt flow is assumed to be in the steady state, $i.e.,$ Q_p=constant.

The following equation may be used to represent the system.

$$\Delta y(t+1) = c_0(t)\Delta u(t) + c_1(t)\Delta u(t-1) + c_2(t)\Delta u(t-2)$$
$$+c_3\Delta u(t-3) + c_4(t)\Delta u(t-4) + c_5(t)\Delta u(t-5)$$
$$+c_6(t)\Delta u(t-6).$$

It can be applied to the controller (7.38) with leakage factor $\varrho = 0.987$, control weighting $r = 1$ and prediction horizon $p = 10$. The simulation results are shown in Fig. 7.3, demonstrating that the trajectories of the system response follow the set-point value satisfactorily.

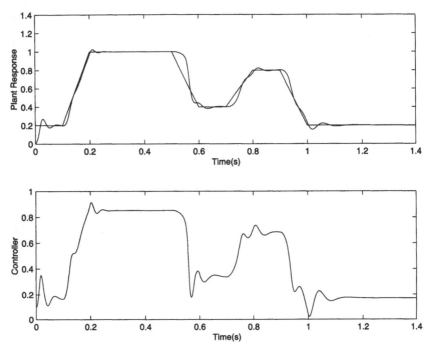

Fig. 7.3. Injection moulding response

7.3.3 Linear Motor Control Problem

Linear motors are becoming very popular for applications requiring high speed, high accuracy operations due to their mechanical simplicity. For these kinds of applications, the control requirements are particularly stringent, which makes the control strategy formulation very challenging.

The linear motor considered here is a brushed permanent magnet DC linear motor produced by Anorad Corp. A non-linear model can be used to describe the dynamics of the linear motor:

$$\dot{x} = v, \tag{7.90}$$

$$\dot{v} = \frac{u - f_{friction} - f_{ripple}}{m} + w(t), \tag{7.91}$$

where $f_{friction}$ is the friction force, f_{ripple} is the ripple force, u is the developed force, m is the combined mass of translator and load, and w represents any other residual disturbances.

The friction and ripple forces are assumed to be modelled as follows:

$$f_{friction} = (f_c + (f_s - f_c)e^{-(\dot{x}/\dot{x}_s)^\delta} + f_v\dot{x})\text{sgn}(\dot{x}), \qquad (7.92)$$
$$f_{ripple} = b_1\sin(w_0 x). \qquad (7.93)$$

The above model allows the evaluation of the friction force during both sticking and slipping motions. The model parameters can be found in Tan *et al.* (1998) and Tan *et al.* (2001).

The desired trajectories are planned as follows:

$$x_d(\tau) = x_0 + (x_0 - x_f)(15\tau^4 - 6\tau^5 - 10\tau^3),$$
$$v_d(\tau) = (x_0 - x_f)(60\tau^3 - 30\tau^4 - 30\tau^2),$$

where $\tau = t/(t_f - t_0)$, x_0 is the initial position and x_f is the final position. In this simulation study, choose $x_0 = 0, x_f = 0.2\text{m}, t_f = 1\text{s}$.

To apply the controller, the system is modelled by

$$\Delta y(t+1) = c_0(t)\Delta u(t) + c_1(t)\Delta u(t-1) + c_2(t)\Delta u(t-2)$$
$$+c_3(t)\Delta u(t-3) + c_4(t)\Delta u(t-4) + c_5(t)\Delta u(t-5).$$

Choose $\varrho = 0.99, p = 10, r = 5$. Figure 7.4 shows the process response, indicating that the control has resulted in satisfactory performance.

Remark 7.4

Since the control law can be applied to a large class of non-linear systems, there is a price to be paid, and that is the amount of control effort required is substantially increased. It is the well-known performance versus control effort dilemma.

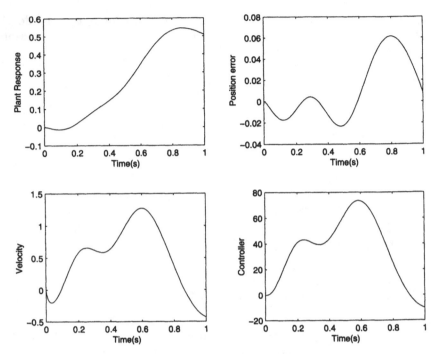

Fig. 7.4. Linear motor response

CHAPTER 8
CASE STUDIES

8.1 Introduction

This chapter is devoted to three case studies involving the application of predictive control systems. A particular purpose of this chapter is to emphasise the key issues involved in the practical application of predictive control to real processes.

Over the last few years, many applications of predictive controllers have been reported. For example, In Gambier and Unbehauen (1991), the application of the generalised predictive control (GPC) to a turbo-generator pilot plant has been described. Application of an extended dynamic matrix control (DMC) to control pH in a neutralisation reactor has been reported in Draeger *et al.* (1995). The application of non-linear model predictive control to a continuous stirred tank reactor (CSTR) has been considered in, for example, Piche *et al.* (2000). These references represent only a small number of the many papers that have been written on the applications of predictive controllers. Many other application papers on the subject can be found in Martin Sanchez and Rodellar (1996).

In this chapter, three case study applications of the predictive controllers are discussed. The first case study concerns the control of an intelligent vehicle. The purpose of the study is to show the methodology of predictive control, such as the choice of the desired trajectory and how to deal with constraint conditions in the control system. This application is examined by running a highway simulator under very realistic traffic assumptions.

The second case shows that the predictive proportional-integral-derivative (PID) controller of Chapter 5 can be used as a position controller for an injection moulding machine. The purpose of this case study is to show that even if a simple model (second-order model) of the process is used, the performance improvements can be significant compared to a conventional PID controller. This case study includes experiments on the servo-hydraulic unit of an injection moulding machine.

The third case study provides results of the application of the adaptive predictive control scheme presented in Chapter 7 to a DC motor. The application shows that the adaptive predictive controller can be used on systems with unknown models.

8.2 Application to Intelligent Autonomous Vehicle

Traffic congestion is a world-wide problem. Historically, this problem is solved by building new highways. Unfortunately, adding highways is not always a viable solution in many areas for a number of reasons: lack of suitable land, escalating construction costs, environmental considerations, *etc.* Because of these and other constraints, different ways to increase capacity must be found. One possible way to improve capacity is to use current highways more efficiently by removing as much human involvement as possible from the system through computer control and automation. Simulation results in Varaiya (1993) show that full automation can accommodate more cars, with an expected fourfold increase in capacity over highways with totally manual traffic. In addition to capacity, automation may make driving and transportation in general safer. In recent years, in Japan, Europe, and the USA, there has been an increased interest in developing automated vehicle technology (Walzer and Zimdahl, 1988; Ito *et al.*, 1990; Bender, 1991). Previous work on the design automated vehicle control systems falls into three categories:

(1) Intelligent cruise control, such as adaptive cruise control, continuous platooning (Ioannou and Chien, 1993). Complex scenarios are not addressed in this work.

(2) Hierarchical control system design for fully automated traffic (Varaiya, 1993; Eskafi *et al.*, 1994). The control system design requires extensive vehicle-to-vehicle coordination.

(3) Fully autonomous vehicle control system based on artificial intelligence (Niehaus and Stengel, 1994). The work in this class is mostly concerned with handling complex traffic scenarios and does not attempt to optimise control performance for vehicle spacing and passenger comfort.

Since driving a vehicle on a limited-access highway requires many different functions to be performed, including traffic situation analysis and careful planning of the following trajectory, as well as vehicle control, it is necessary to design any decision and control system from a safety viewpoint when considering automated vehicles (Huang and Ren, 1998).

This section describes the design and experimental evaluation of a completely self-contained intelligent vehicle controller, called the autonomous vehicle driving system (AVDS), for automatic vehicle driving on a limited-access highway. AVDS is a two-level structure: the top level, called the decision level, analyses the traffic situation and determines a driving mode among several basic driving modes, while the bottom level uses predictive control to follow different driving modes. One of the salient features of the hierarchical controller is that the bottom level of the structure utilises predictive control to optimise the control objective provided by the top level, while considering passenger comfort and safety. In this way, the controller at the bottom level has the same structure across different driving modes.

8.2.1 Description of Intelligent Vehicle Control

In a typical envisioned driving scenario, the human driver first guides the car into a highway traffic flow, and then sets the path of the vehicle, such as lane number and exit name, to the on-board computer. After this setting is completed, the driver presses a button to hand over control to the computer and the vehicle starts automatic control by use of the on-board computer and its sensors.

Assume that the sensors on the automated vehicles have the following capabilities:

- They can provide velocity, and distance measurements of the vehicle in front, as well as measurements of the vehicle's own velocity and acceleration.

- They can detect vehicles in adjacent lanes (see Fig. 8.1). The detection ranges are taken to be constants.

  - In region 1 and 2, 60 metres.

  - In region 3 and 4, ± 30 metres from the centre of the car.

- They can detect signals from lane number, highway number and exit markers.

- They can detect the changing-lane light sent from other vehicles in adjacent lanes.

- They can receive the optimal speed and safe spacing parameters provided by the roadway centre (this can be implemented by measuring the message signs placed on the road).

With these assumptions in mind, the automated vehicles should handle all possible traffic scenarios and try to retain their assigned lane unless particular

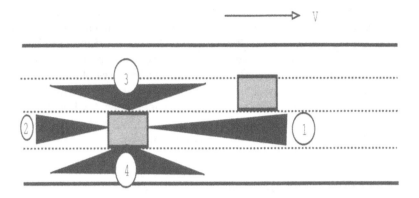

Fig. 8.1. Detection ranges of sensors

situations occur, such as emergency or preceding vehicle too slow, *etc.* A two-level hierarchical structure is designed for the AVDS (see Fig. 8.2). The top level, called the decision level, reads the information coming from the roadside centre and on-board sensors and determines an appropriate driving mode among a set of basic driving modes, as well as the associated reference trajectory. The roadside centre is a traffic service centre for one branch of a highway network. It receives traffic flow measurements from the sections and broadcasts optimal speed and headway spacing to the vehicles in its section. The bottom level, called the control level, issues control commands to the vehicle. This control scheme is in agreement with the intelligent hierarchy principle – that there is "increasing intelligence with decreasing precision as one moves from the lower to the higher levels". Thus, it is possible that complex behaviour may be more easily achieved with intelligent inference, while the basic control task may be implemented by a precise algorithm.

An inference system at the top level is used to make decisions and to give the correct driving mode. The decision must satisfy safety criteria and provide an appropriate trajectory for vehicle driving. Once the objective has been identified, the decision level results will be transmitted to the lower level.

Based on the reference commands instructed by the top level, a predictive controller at the bottom level implements the requirements of all modes. This controller guarantee the safety and ride comfort of the passengers.

To efficiently connect the decision (driving mode) and the control (algorithm) level, an interface is used in the AVDS system, which can transfer the result of the decision at the top level into a simple vector form suitable for the control algorithm at the bottom level. Such a design for AVDS is more modular and extensible in the sense that new rules can be added in the top level, or the structure of the prediction controller can be changed in the bottom level. The detailed design of each level will be shown in the following sections.

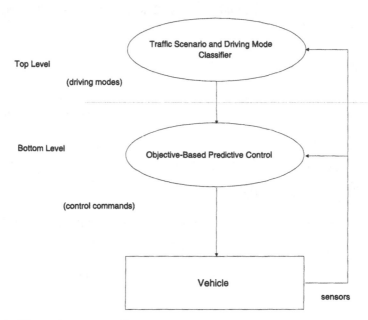

Fig. 8.2. Hierarchical controller

8.2.2 Decision-making

In this section, discussions will focus on how to make decisions when facing complex traffic scenarios. As a prelude to the problem, the basic driving modes and associated reference trajectories are first defined. The goal of the driving modes chosen is to ensure safety, and to achieve a compromise between control performance and passenger comfort.

The reference trajectory $w(k)$ corresponding to each driving mode takes the form

$$w(k+1) = \eta w_d(k) + (1 - \eta)y(t), \tag{8.1}$$

where $y(t)$ is the measured variable, w_d is the output set-point, and $0 < \eta < 1$ for a slow transition from the current measured-variable to the output set-point $w_d(k)$ (De Keyser *et al.*, 1988). In general, $\eta = e^{-\tau/T}$ where τ is the time constant and T is the sampling instant which is defined in (8.15).

Velocity tracking mode. In this mode, the control tries to force the vehicle to track an optimal velocity.

Let v_{opt} be the desired optimal velocity. The reference velocity trajectory then takes the form

$$w_v(k+1) = \eta v_{opt} + (1-\eta)v(k), \tag{8.2}$$

where the parameter η is used to adjust the aggressiveness of the control. To see the reason for choosing such a control objective, consider the ideal situation

$$v(k+1) - v_{opt}\eta - (1-\eta)v(k) = 0, \quad k \geq 0, \tag{8.3}$$
$$v(k+1) = v_{opt} + [v(0) - v_{opt}](1-\eta)^{k+1}. \tag{8.4}$$

This defines a smooth reference trajectory from the initial velocity $v(0)$ to the desired velocity v_{opt}. The reference trajectory is plotted below for different η (Fig. 8.3). Notice that smaller values of η give more passenger comfort. Usually, this mode is operated within a safe distance. Let $\eta = e^{-\frac{0.05}{T}}$ in this mode.

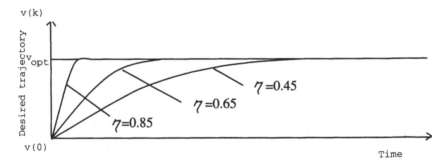

Fig. 8.3. Desired trajectory for different values of η

Velocity keeping mode. In this mode, the control objective is simply to maintain vehicle velocity, and the reference velocity trajectory is

$$w_v(k+1) = \eta v(k_0) + (1-\eta)v(k), \tag{8.5}$$

where $v(k_0)$ is the vehicle speed at time $t = k_0$, and k_0 is the initial time of this mode. The parameter η has the same selection as the above.

Weak spacing control mode. In this mode, the control forces the vehicle to smoothly follow the vehicle immediately in front of it with a certain desired spacing.

Let s_d be the desired spacing. The reference spacing trajectory takes the form

$$w_s(k+1) = \eta s_d + (1 - \eta)s(k), \tag{8.6}$$

where one can again choose a suitable η to ensure a smooth trajectory. Set $\eta = e^{-\frac{0.048}{T}}$ in this mode.

Strong spacing control mode. In this mode, the control should force the vehicle to rapidly follow the vehicle immediately in front of it with a certain desired spacing. The reference trajectory takes the same form as suggested by (8.6). However, a larger value of η is set, that is $\eta = e^{-\frac{0.025}{T}}$ in this mode.

Emergency mode. In this mode, due to emergency or near-emergency situations, the control is required to assess safety by predicting future trajectories, to decide whether to change lanes, to plan a trajectory, or to execute the trajectory. The reference trajectory takes the form

$$w_s(k+1) = \eta s_d^E(k_0) + (1 - \eta)s(k), \tag{8.7}$$

where $s_d^E(k_0)$ is the vehicle's distance at time $t = k_0$, and k_0 is the initial time of this mode. The recommended value is $\eta = -\frac{0.02}{T}$ in this mode.

Boundary control mode. This mode requires the vehicle to attempt maximum deceleration in order to avoid a collision.

Creating mode. This mode requires the vehicle to create a space so that a WANT-TO-CHANGE-LANE vehicle in an adjacent lane can change lanes to be in front of it. The design should force the vehicle to engage weak spacing control. The reference trajectory takes the form

$$w_s(k+1) = \eta s_d^C + (1 - \eta)s_L(k), \tag{8.8}$$

where $s_d^C(k)$ is the space requirement from the WANT-TO-CHANGE-LANE vehicle (typically 30m) and $s_L(k)$ is the longitudinal distance between itself and the WANT-TO-CHANGE-LANE vehicle. The recommended value of η is $\eta = -\frac{0.056}{T}$ in this mode.

Change lane mode. This mode moves the vehicle to the adjacent lane. It requires mainly lateral movement.

To determine an appropriate driving mode, the workings of the decision are usually similar to human driver behaviour. Figures 8.4 and 8.5 describe the flow diagram for the automated driving decision at the top level, where v is the velocity of the vehicle controlled, v_h is the velocity of the preceding vehicle, v_{opt} is the desired velocity, s is the actual spacing from the preceding vehicle, and s_d is the desired spacing. At every iteration of the control loop, the decision begins from the Top Level Executive. The decision is also activated when it receives an interrupt signal from the interface (this will be further discussed in the sub-section on the interface). The driving mode can be obtained by checking the conditions marked in the ellipse blocks along the logic tree in Figs 8.4 and 8.5. Each condition is written as a parameter $(True, False, ConditionName)$. For example, if the driving mode obtained by the last control loop is *emergency*, then the parameter of *check if emergency is* $(True, Nil, Emergency)$; if the on-board sensor detects an exit beacon, then the parameter of *check if exiting flag* is $(True, Nil, ExitingFlag)$. To help understand the decision-tree flow diagram, optional conditions in the ellipse blocks need to be explained clearly:

- *check if exiting flag* is a flag indicating that the vehicle will exit from the highway. This flag set by the computer when the on-board sensor detects an exit beacon on the roadside and it coincides with the setting exit name;

- *check if it is unsafe* examines the signal of the Safety Flag 1 in the interface (see the definition of Safety Flag 1 in the discussion section on the interface). If the Safety Flag 1 signifies success, the parameter of this condition is $(Nil, False, Unsafe)$. Otherwise, the parameter of the condition is $(True, Nil, Unsafe)$;

- *check a space in adjacent lane* is used to check if there is a 'safe' space (typically 30m) in an adjacent lane to accommodate merging. The meaning of safe space here implies that there is an appropriate space and the on-board sensor does not 'see' a change-lane-light in its sensor range. This prevents two vehicles merging into the same spot;

- *check if change-lane flag* is a flag indicating that vehicle is in a lane-change mode;

- *check if it has completed lane change* determines if the vehicle has moved into a new lane;

- *check if it is in its assigned lane* determines if the vehicle is in its assigned lane that was set in advance by the human driver. If *yes*, then the vehicle will search the next branch down in the decision tree. Otherwise, the vehicle will try to move to the adjacent lane closest to the assigned lane;

Fig. 8.4. Driving mode decision (1)

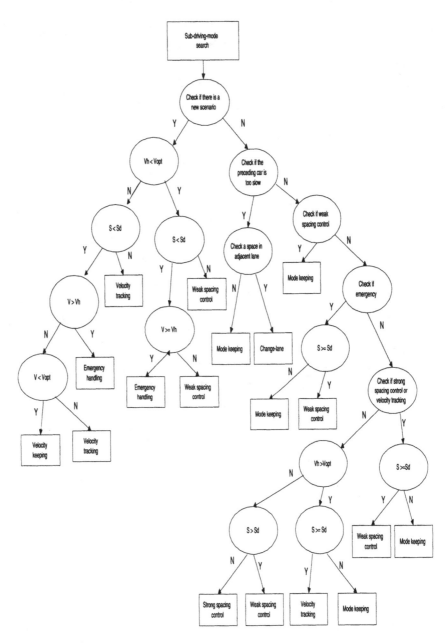

Fig. 8.5. Driving mode decision (2)

- *lane-change request from another vehicle* checks if the changing-lane light sent from another vehicle in an adjacent lane has been detected. This case will be discussed later;

- *check if it is a closest car* determines if the vehicle is closest to a WANT-TO-LANE-CHANGE vehicle in an adjacent lane;

- *check if there is a car ahead* determines if the car using its own sensors can "see" a car in front;

- *check if there is a new scenario* determines if the controlled vehicle encounters a new scenario in front. It includes three cases (see Fig. 8.6):

  - case 1 in Fig. 8.6. Assume that the current time is $t = k$. The controlled car, called A, does not detect any car along the motion direction at time $t = k - 1$. At time $t = k$, it sees a Car B ahead;

  - case 2 in Fig. 8.6. Assume that the current time is $t = k$. The controlled car A follows Car C and does not find Car B in its current lane at time $t = k - 1$. At time $t = k$, it detects Car B changing lane from lane 2 into its current lane 1;

  - case 3 in Fig. 8.6. Assume that the current time is $t = k$. The controlled car A stays in lane 2 at time $t = k - 1$. At time $t = k$, Car A moves in lane 1 and finds Car B in front;

- *check if the preceding car is too slow* determines if the speed v_h of the immediate preceding car is far less than the desired speed v_{opt}. If $v_h < v_d - 10\text{ms}^{-1}$, then the immediate preceding car is too slow;

- *check if weak spacing control (emergency, strong spacing control, or velocity tracking)* examines what driving mode was in operation at the last sampling instant;

- *mode keeping* maintains the same mode as the result obtained at the decision level's last sampling instant.

The other optional conditions are clear. By using these conditions about traffic situations, the goal of the decision is to infer the driving mode along the logic tree as shown in Figs 8.4 and 8.5. Consider a driving example (Fig. 8.4). First, the system determines if there is an exiting flag. If *no*, and if the car does not show the changing-lane flag, its own lane is examined. If the system recognises that the current lane coincides with the assigned lane set by the human driver, then the sensors will examine if there is a changing-lane light sent from an adjacent vehicle nearby. When WANT-TO-CHANGE-LANE light is detected, and the car is closest to that car, the decision will occur. When no changing-lane light is detected, the system determines if there is a

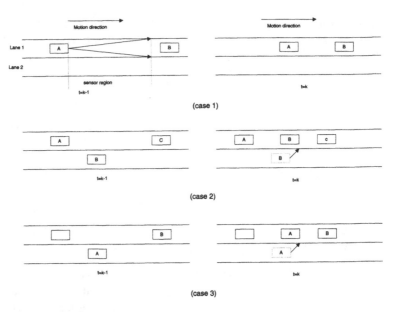

Fig. 8.6. There is a new scenario

car ahead in the same lane. If *yes*, the system enters the sub-driving-mode search tree to deduce an appropriate driving mode. When there is no car ahead, the system issues *velocity tracking mode*. At every control loop, the decision has to reach a driving mode and notify the interface to execute this mode.

8.2.3 Interface between Decision Level and Control Level

As discussed above, the decision level modes are typically of mixed form containing a symbol and numeric expression, such as "velocity tracking mode" with the associated reference trajectory $w_v(k+1) = \eta v_{opt} + (1 - \eta)v(k)$. The prediction control algorithm at the bottom level is unable to directly interpret these modes and introduce them into the actuator input calculations. Similarly, the decision level needs some way of interpreting the result of the calculations at the bottom level in a form that it can understand. In this section it is intended to seek to provide an interface that allows them to exchange messages.

Vector

$$(mode\_type(\theta), safety\ flag\ 1, safety\ flag\ 2)$$

is used to represent this interface, where *mode_type* includes velocity keeping, velocity tracking, weak spacing, strong spacing, emergency and change lane, $\theta = (s_d \text{ or } v_d, \eta)$ is the associated parameter, Safety Flag 1 is used to notify the decision level if the control required for a particular driving mode is safe through estimates of the prediction algorithm at the bottom level, and Safety Flag 2 is an interrupt line used to re-activate the decision level when Safety Flag 1 shows an unsafe signal. Safety Flag 2 depends on the message passed by Safety Flag 1.

The decision is involved in the interface in some ways. At each control loop, the decision level has to generate a driving mode with its responding reference trajectory to the interface. The interface reads the mode and calls the control level to carry it out. Based on the mode type and reference trajectory, the prediction algorithm at the bottom level calculates the actuator input needed to drive the vehicle as close as possible to the desired trajectory. In a normal traffic situation, the algorithm is completed without affecting safety constraints, *i.e.*, hardware limits and minimum separation distance, thereby resulting in a safe control input for vehicle driving. In this case, the interface stores a flag that signifies success in Safety Flag 1. This flag is used by the decision analysis on the next iteration. Safety Flag 2 (interrupt line) is not triggered to activate the decision. In an emergency, however, there are two possible calculation results obtained by the control level: one is that the prediction algorithm, without affecting the safety constraints, is completed; the other is that the prediction algorithm is aborted due to violation of safety constraints. If the former happens, the interface has the same actions as in a normal traffic situation. If the latter happens, this means that control under the current scenario and hardware technology cannot guarantee safety. The Safety Flag 1 signifies a failure signal and no control value is available. In this case, Safety Flag 2 (interrupt line) activates the decision level. The decision begins again from *Top Level Executive* of the logic tree in Fig. 8.4. Since *check if it is unsafe* in the logic tree has become *true* due to the Safety Flag 1 signifying failure, the system turns to *check a space in adjacent lane*. If there is enough space in an adjacent lane, the change-lane mode is entered. Otherwise, the boundary control mode is detected. Unlike the other driving modes, the change-lane or boundary control is a definite command and the control level does not need to use the prediction algorithm to estimate the actuator input. Thus, the interface notifies the control level executing the command and signifies success in the Safety Flag 1. The control level sends the command direct to the actuator. At the next sampling instant, the AVDS system begins again.

In summary, when the vehicle is in a normal situation the decision is executed once every sampling period. However, when the vehicle is in an emergency situation it is likely to be executed twice every sampling period until the traffic scenario is safe.

8.2.4 Optimal Tracking Based on Predictive Control

Below the decision level in the system hierarchy lies the control level. Its task is to implement the driving mode instructed by the higher level. For this purpose it uses a control algorithm to calculate the actuator input for a particular driving mode. In this section, a predictive control algorithm is used to implement this.

The closed-loop behaviour of predictive control algorithms depends primarily on the output "prediction horizon" which defines the number of future sampled outputs to be considered in the cost function minimisation, and the "control horizon" which specifies the number of future control moves needed to achieve control objective. Naturally, such a look-ahead control is a safe method for vehicle driving.

Consider the following simplified non-linear vehicle model:

$$\dot{x} = v, \tag{8.9}$$

$$\dot{v} = -\frac{1}{m}(F_w - F_e + F_d), \tag{8.10}$$

where x is the longitudinal position of the vehicle, v is the velocity of the vehicle, m is the mass of the vehicle, $F_w = K_w v^2$ is the force due to wind resistance, where K_w is the aerodynamic drag coefficient, F_d is the force due to mechanical drag, and F_e is the engine traction force, which is assumed to evolve under the dynamics:

$$\dot{F}_e = -\frac{F_e}{\tau} + \frac{u}{\tau}, \tag{8.11}$$

where τ is the engine time constant and u is the throttle/brake input. Define $a = \dot{v}$; it is found that

$$\dot{a} = -\frac{1}{m}(2K_w v\dot{v} + \frac{F_e}{\tau} - \frac{u}{\tau}).$$

Rewriting F_e using the equation (8.10) above,

$$F_e = ma + K_w v^2 + F_d,$$

it follows that

$$\dot{a} = -\frac{1}{m\tau}[2\tau K_w v\dot{v} + ma + K_w v^2 + F_d - u].$$

Under the assumptions that both the velocity and acceleration of the vehicle and the other vehicle parameters are available, the model can be linearised by state feedback. Setting

$$u = m\tau(c + 2\tau vaK_w + K_w v^2 + F_d),$$

the new linearised vehicle model is given by $\dot{a} = c$, where c is an exogenous control input. Now the dynamics of the linearised vehicle are summarised by the three simple state equations:

$$\dot{x} = v, \tag{8.12}$$
$$\dot{v} = a, \tag{8.13}$$
$$\dot{a} = c. \tag{8.14}$$

Since the algorithm is a discrete-time predictive control, the discrete motion equation can be described as

$$x(k+1) = x(k) + v(k)T + \frac{1}{2}a(k)T^2, \tag{8.15}$$
$$v(k+1) = v(k) + a(k)T, \tag{8.16}$$
$$a(k+1) = a(k) + c(k)T, \tag{8.17}$$

where T is the sampling instant.

For each vehicle, the relative distance between itself and the immediately preceding vehicle is defined by

$$s(k) = x_p(k) - x(k), \tag{8.18}$$

where $x(k)$ and $x_p(k)$ are the positions of itself and the immediate preceeding vehicle, respectively, at $t = k$. From the motion equation (8.15), this is given by

$$s(k+1) = s(k) - v(k)T - \frac{1}{2}a(k)T^2 + v_p(k)T + \frac{1}{2}a_p(k)T^2. \tag{8.19}$$

Combining (8.16) and (8.17) yields

$$\begin{pmatrix} s(k+1) \\ v(k+1) \\ a(k+1) \end{pmatrix} = \begin{pmatrix} 1 & -T & -0.5T^2 \\ 0 & 1 & T \\ 0 & 0 & 1 \end{pmatrix} \begin{pmatrix} s(k) \\ v(k) \\ a(k) \end{pmatrix} + \begin{pmatrix} 0 \\ 0 \\ T \end{pmatrix} c(k)$$
$$+ \begin{pmatrix} T \\ 0 \\ 0 \end{pmatrix} (v_p(k) + \frac{1}{2}a_p(k)T). \tag{8.20}$$

Let

$$z(k) = [s(k) \quad v(k) \quad a(k)]^T,$$

$$A = \begin{pmatrix} 1 & -T & -0.5T^2 \\ 0 & 1 & T \\ 0 & 0 & 1 \end{pmatrix}, \quad B = \begin{pmatrix} 0 \\ 0 \\ T \end{pmatrix}, \quad C = \begin{pmatrix} T \\ 0 \\ 0 \end{pmatrix},$$

$$D = \begin{cases} (1 \ 0 \ 0) & \text{for the spacing control} \\ (0 \ 1 \ 0) & \text{for the velocity control} \end{cases}$$

Thus,

$$z(k+1) = Az(k) + Bc(k) + C(v_p(k) + \frac{1}{2}a_p(k)T), \tag{8.21}$$

$$y(k) = Dz(k). \tag{8.22}$$

The structure of the model (8.21) is useful in the formulation of a predictive controller. Based on the model (8.21), a state prediction model is defined as

$$\hat{z}(k+j) = A\hat{z}(k+j-1) + Bc(k+j-1) + Cv_p(k+j-1); $$
$$j = 1, 2, ..., P, \tag{8.23}$$

where $\hat{z}(k+j)$ denotes the state vector prediction at instant k for instant $k+j$ and $c(k+j-1)$ denotes the sequence of control vectors on the prediction interval. This model is refined at each sampling instant k from the actual state vector

$$\hat{z}(k) = z(k). \tag{8.24}$$

Note that the prediction model (8.23) does not include $a_p(k)$, since it is difficult to detect the acceleration of the immediately preceding vehicle. This will result in a model error. Fortunately, the error can be reduced significantly since the prediction model is updated at each sampling instant.

At each time $t = k$, a sequence (length M) of controls $c(k), ..., c(k+M-1)$ is selected so that the quadratic cost function

$$J = \sum_{l=1}^{P} \|w(k+l) - \hat{y}(k+l)\|_{Q(l)}^2 + \sum_{j=1}^{M} \|c(k+j-1)\|_{R(j)}^2, \tag{8.25}$$

is minimised under the constraints (8.21) and (8.22), where $w(k+l)$ is the desired value of $\hat{y}$ at $t = k, Q(l) \geq 0$ is the output-error weighting matrix,

$R(j) \geq 0$ is the control weighting matrix, and P and M ($N \geq P \geq M$) are called the optimisation and control horizon, respectively.

In order to obtain the future outputs $\hat{y}(k+l)$ in (8.22), a prediction with this control sequence is performed on the basis of the predictive model (8.22):

$$\begin{aligned}
\hat{y}(k+1) &= DAz(k) + DBc(k) + DCv_p(k), \\
\hat{y}(k+2) &= DA^2z(k) + DABc(k) + DBc(k+1) \\
&\quad + (DAC + DC)v_p(k),
\end{aligned}$$

......

$$\begin{aligned}
\hat{y}(k+P) &= DA^P z(k) + DA^{P-1}Bc(k) + ... + DA^{P-M}Bc(k+M-1) \\
&\quad + (DA^{P-1} + DA^{P-2} + ... + D)Cv_p(k),
\end{aligned}$$

where the speed of the preceding vehicle is assumed to be given by $v_p(k) = v_p(k+1) = ... = v_p(k+P)$ over the prediction interval. The equations above can also be rewritten as

$$Y = EAz(k) + FU + Gv_p(k), \tag{8.26}$$

where

$$Y = [\hat{y}(k+1)...\hat{y}(k+P)]^T, \tag{8.27}$$

$$U = [c(k)...c(k+M-1)]^T, \tag{8.28}$$

$$E = \begin{pmatrix} D \\ DA \\ \vdots \\ DA^{P-1} \end{pmatrix}, \tag{8.29}$$

$$F = \begin{pmatrix} DB & 0 & 0 & 0 \\ DAB & DB & 0 & 0 \\ \vdots & & \ddots & \\ DA^{P-1}B & DA^{P-2}B & ... & DA^{P-M}B \end{pmatrix}, \tag{8.30}$$

$$G = \begin{pmatrix} DC \\ DAC + DC \\ \vdots \\ DA^{P-1}C + DA^{P-2}C + ...DC \end{pmatrix}. \tag{8.31}$$

Substituting (8.26) into (8.25), one obtains

$$\begin{aligned}
J &= (W - EAz(k) - FU - Gv_p(k))^T Q(W - EAz(k) - FU - Gv_p(k)) \\
&\quad + U^T RU,
\end{aligned} \tag{8.32}$$

where

$$W = [w(k+1), w(k+2), ..., w(k+P)]^T, \qquad (8.33)$$

$$Q = Blockdiag\{Q(1), ..., Q(P)\} \geq 0, \qquad (8.34)$$

$$R = Blockdiag\{R(1), ..., R(M)\} \geq 0. \qquad (8.35)$$

Since vehicles are normally affected by limitations, the predictive control should have the capability to account for such constraints. Failure to do so could result in uncomfortable passengers, and even instability. A quadratic programming (QP) approach may be used to solve the constrained cost function (Garcia and Morshedi, 1984). Inspection of (8.32) shows that the optimisation problem being considered is least squares with linear inequality constraints

$$\min U^T H U + LU \qquad (8.36)$$

$$\text{subject to } SU \leq f, \qquad (8.37)$$

where $H = R + F^T QF$, $L = 2[z^T(k)A^T E^T + v_p^T(k)G^T - W^T]QF$. The matrix S contains dynamic information about the constraints while f gives the limiting values for the constraints.

For this design several constraint considerations for vehicle operation are imposed. The first is ride quality for the passengers. It is well known that ride comfort requires small acceleration and jerks. The required limitations on jerks and accelerations to achieve ride comfort may be taken as $[-5, 5]\text{ms}^{-3}$ and $[-5, 2]\text{ms}^{-2}$, respectively.

The car should try to maintain this comfort level whatever situation it is in. However, the constraints above may be violated when there is a safety problem. In this case, instead of given hardware limits, jerks and accelerations are taken as $[-30, 10]\text{ms}^3$ and $[-15, 5]\text{ms}^2$ for jerks and accelerations, respectively. If the hardware limits are violated, this implies that the vehicle may not be able to track the determined driving mode provided by the decision level under the current hardware technology. In this case, the car staying in its own lane has a higher possibility of collision, and so it should try to change lanes. Standard forms for jerk and acceleration constraints with the M control horizon are given by

$$S_1 = \begin{pmatrix} 1 & 0 & 0 & \dots & 0 & 0 & 0 \\ -1 & 0 & 0 & \dots & 0 & 0 & 0 \\ 0 & 1 & 0 & \dots & 0 & 0 & 0 \\ 0 & -1 & 0 & \dots & 0 & 0 & 0 \\ \cdot & \cdot & \cdot & \cdot & \cdot & \cdot & \cdot \\ 0 & 0 & 0 & \dots & 0 & 0 & 1 \\ 0 & 0 & 0 & \dots & 0 & 0 & -1 \end{pmatrix}_{2M \times M},$$

$$f_1 = \begin{pmatrix} j_{max} \\ j_{min} \\ \vdots \\ j_{max} \\ j_{min} \end{pmatrix}_{2M},$$

$$S_1 U \leq f_1, \tag{8.38}$$

where $[j_{min}, j_{max}]$ are either ride comfort constraints or hardware constraints, which depend on the requirements of the driving mode.

The acceleration constraints with the M control horizon are:

$$S_2 = \begin{pmatrix} 1 & 0 & 0 & \dots & 0 & 0 & 0 \\ -1 & 0 & 0 & \dots & 0 & 0 & 0 \\ 1 & 1 & 0 & \dots & 0 & 0 & 0 \\ -1 & -1 & 0 & \dots & 0 & 0 & 0 \\ \cdot & \cdot & \cdot & \cdot & \cdot & \cdot & \cdot \\ 1 & 1 & 1 & \dots & 1 & 1 & 1 \\ -1 & -1 & -1 & \dots & -1 & -1 & -1 \end{pmatrix}_{2M \times M},$$

$$f_2 = \begin{pmatrix} \frac{a_{max} - a(k)}{T} \\ \frac{-a_{min} + a(k)}{T} \\ \vdots \\ \frac{a_{max} - a(k)}{T} \\ \frac{-a_{min} + a(k)}{T} \end{pmatrix}_{2M},$$

$$S_2 U \leq f_2, \tag{8.39}$$

where $[a_{min}, a_{max}]$ have similar choices to the jerk constraints above.

For the output constraints the predictive spacing of the controlled car should be greater than the minimum separation distance d_0 (typically 0.5m). Let y_s

represent the spacing for the vehicle. From (8.22),

$$y_s(k) = Dz(k), \tag{8.40}$$

where $D = (1 \ 0 \ 0)$. Similar to (8.26), one obtains for the predictive spacing vector of the controlled car

$$Y_s = EAz(k) + FU + Gv_p(k) \geq \begin{pmatrix} d_0 \\ \vdots \\ d_0 \end{pmatrix},$$

which implies that

$$-FU = S_3U \leq EAz(k) + Gv_p(k) - \begin{pmatrix} d_0 \\ \vdots \\ d_0 \end{pmatrix} = f_3,$$

$$S_3U \leq f_3, \tag{8.41}$$

where $Y_s = [\hat{y}_s(k+1)...\hat{y}_s(k+P)]^T$.

In this section, the conditions (8.38)–(8.41) are referred to as safety constraints. The vehicle should try to avoid violating these constraints.

Based on the output reference trajectory $w(k+1)$, the task of the prediction algorithm is to solve the optimisation problem (8.36) subject to the safety constraints, obtaining the actuator input value. How the algorithm at the control level executes the interface commands is now explained.

- When "mode_type" is *velocity tracking mode* with the corresponding parameter $\theta = (v_{opt}, e^{-\frac{0.05}{T}})$, the reference trajectory is written as

$$w(k+1) = e^{-\frac{0.05}{T}} v_{opt} + (1 - e^{-\frac{0.05}{T}})v(k). \tag{8.42}$$

Based on this objective, control can be obtained by optimizing (8.36) subject to the condition

$$\begin{pmatrix} S_1 \\ S_2 \end{pmatrix} U \leq \begin{pmatrix} f_1 \\ f_2 \end{pmatrix}, \tag{8.43}$$

where f_1 and f_2 take the passenger comfort constraint. Note that it is not necessary to consider the mimimum separation distance constraint in this mode .

- When "mode-type" is *velocity keeping mode* with the corresponding parameter $\theta = (v(k_0), e^{-\frac{0.05}{T}})$, the reference trajectory is written as

$$w(k+1) = e^{-\frac{0.05}{T}} v(k_0) + (1 - e^{-\frac{0.05}{T}})v(k). \qquad (8.44)$$

The predictive control is the same as the mode above.

- When "mode-type" is *weak spacing control* with the corresponding parameter $\theta = (s_d, e^{-\frac{0.048}{T}})$, the reference trajectory is written as

$$w(k+1) = e^{-\frac{0.048}{T}} s_d + (1 - e^{-\frac{0.048}{T}})s(k), \qquad (8.45)$$

where s_d is given in (8.6). The actuator input is obtained by optimizing (8.36) subject to

$$\begin{pmatrix} S_1 \\ S_2 \\ S_3 \end{pmatrix} U \le \begin{pmatrix} f_1 \\ f_2 \\ f_3 \end{pmatrix}, \qquad (8.46)$$

where f_1 and f_2 are the passenger comfort constraints.

- When "mode-type" is *strong spacing control* with the corresponding parameter $\theta = (s_d, e^{-\frac{0.025}{T}})$, the reference trajectory is written as

$$w(k+1) = e^{-\frac{0.025}{T}} s_d + (1 - e^{-\frac{0.025}{T}})s(k). \qquad (8.47)$$

The control is similar to the mode above but f_1 and f_2 are the hardware limits.

- When "mode-type" is *creating mode* with the corresponding parameter $\theta = (s_d^c, e^{-\frac{0.056}{T}})$, the reference trajectory for the vehicle is written as

$$w(k+1) = e^{-\frac{0.056}{T}} s_d^c + (1 - e^{-\frac{0.056}{T}})s_L(k), \qquad (8.48)$$

where s_d^c is given in (8.8), and $s_L(k)$ is the relative longitudinal distance between itself and the WANT-TO-CHANGE-LANE car which is in an adjacent lane. The constraints take the same form as for *velocity tracking mode*.

To demonstrate this mode clearly, see Fig. 8.7. Suppose Car A is in Lane 1 and intends to change to Lane 2 (see Fig. 8.7a). It can do so only if there is an adequate space in Lane 2. Car B, in Lane 2, has the closest position to Car A. Thus, the AVDS in Car B decides to stop tracking the car ahead. Instead, it uses *creating mode* to create a space in front of it in order to accommodate A's change lane. The objective is to achieve the desired spacing s_d^c (typically 30m) by using the optimisation calculation (8.36) subject to the constraints. Unlike the usual driving modes, the initial spacing of this mode may take a negative value (see Fig. 8.7b).

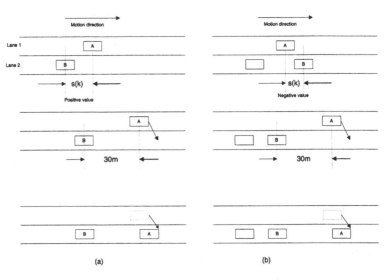

Fig. 8.7. Creating a space for a lane-change car in the right lane

- When "mode_type" is *emergency mode* with the corresponding parameter $\theta = (s_d^E, e^{-\frac{0.02}{T}})$, the reference trajectory is written as

$$w(k+1) = e^{-\frac{0.02}{T}} s_d^E + (1 - e^{-\frac{0.02}{T}}) s(k). \tag{8.49}$$

Solving the optimisation problem (8.36) subject to the constraints is the same as *strong spacing control*. However, safety cannot be guaranteed in the particular situation. The control level will check if the optimisation algorithm has a feasible solution without violating the safety constraints. If *yes*, the control level notifies the interface signifying success in the Safety Flag 1. In this case, control is available. When the algorithm violates the safety constraints, the control level notifies the interface signifying failure in the Safety Flag 1. In this case, the control level waits for the new command from the interface.

- When "mode_type" is *boundary control mode*, the control level applies

$$c(k) = \frac{a_{maxbraking} - a(k)}{T}, \tag{8.50}$$

to the actuator directly, where $a_{maxbraking}$ is the value of the maximum braking acceleration.

- When "mode_type" is *lane change mode*, the simulation software will move the car into an adjacent lane while maintaining speed.

Using these types of reference trajectories, the optimisation problem (8.36) subject to the selected constraints can be solved at every sampling interval. When the prediction algorithm, without affecting the safety constraints, is completed, the control level sends a success signal to the interface, and notifies the actuator executing the control value. When the prediction algorithm violates the safety constraints, the control level sends a failure signal to the interface, and waits for a new command from the interface. Experiments show that only when the driving mode is emergency is the algorithm likely to violate the safety constraints.

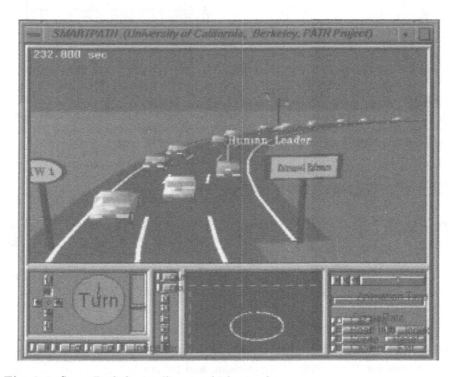

Fig. 8.8. SmartPath for intelligent vehicle simulator

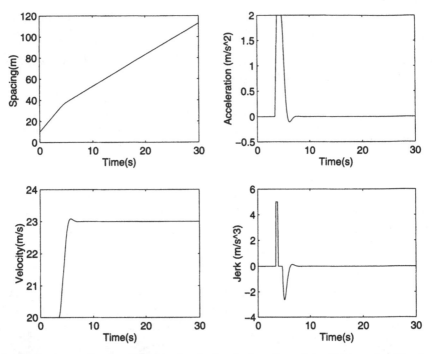

Fig. 8.9. A car changes lanes in front of reducing the safety distance by 10m but that car is going faster than the AVDS-car by 6ms^{-1}. Thus the AVDS-car keeps its original velocity (*velocity keeping mode*) until the spacing is satisfied. Then, it tracks the optimal velocity of 23ms^{-1} (*velocity tracking mode*).

8.2.5 Simulations on SmartPath Simulator

The AVDS described in the previous sections is a complete control system. Based on a simulation package, called SmartPath (Eskafi *et al.*, 1994), the AVDS design has been successfully implemented on a Silicon Graphics workstation. SmartPath is an AHS simulator package (see Fig. 8.8). The program may be used to understand how the AHS would perform in terms of highway capacity, traffic flow, and other performance measures of interest to transportation system planners and to drivers.

In this section, some simulation studies are conducted for the proposed system which combine the driving mode decision and the predictive control. Assume that the pavement is dry. Constants $K_w = 0.44$kgm^{-1}, m=916kg,$\tau = 0.2$s and $F_d = 352$kgms^{-2} are used in the simuation. The sampling instant is taken as T=0.08s. During the simulations, there are two tuning parameters to be considered: the prediction horizon P, and the control weighting matrix R. By experiment, the length of the predicted output for P is chosen as 15 while the length of the input horizon is $M = 15$. The control weighting matrix for R is

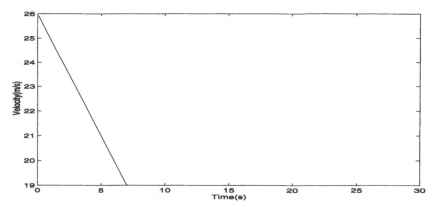

Fig. 8.10. Velocity profile of the cut-in vehicle

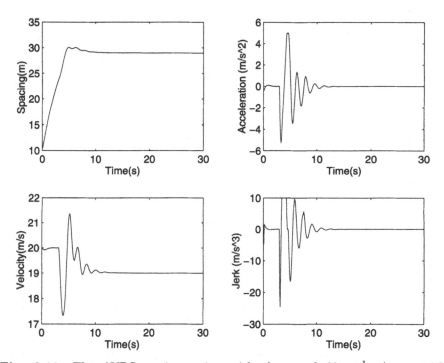

Fig. 8.11. The AVDS-car is moving with the speed 20ms^{-1}. A car with speed 26ms^{-1} and acceleration -1ms^{-2} changes lanes in front of it reducing the safety distance by 10m. Initially, the AVDS-car keeps its original velocity (*velocity keeping mode*) until that car's speed is less than the optimum velocity 23ms^{-1}. The AVDS-car uses *strong spacing control* to achieve the safety spacing.

selected as $0.001I_{10\times10}$ while the weighting matrix for Q is $I_{20\times20}$ (where I is the unit matrix).

First, the control performance of the AVDS is illustrated by several examples. The safe spacing is defined in the following way:

$$s_d = \lambda_v v(k-1) + \lambda_p, \tag{8.51}$$

where $v(k-1)$ is the velocity of the car, $\lambda_v = 1\ s$, and $\lambda_p = 10m$. This is a headway spacing dependent on the vehicle's velocity. The optimum velocity v_{opt} is $23\mathrm{ms}^{-1}$. The limitations of jerk and acceleration are given in the fifth section. The plots correspond to the following cases:

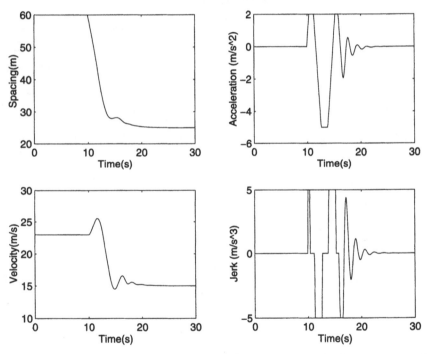

Fig. 8.12. The AVDS-car with speed $23\mathrm{ms}^{-1}$ has no cars in its distance sensor for $t < 10\mathrm{s}$. At $t=10\mathrm{s}$, it sees a car $60m$ ahead going at a speed $15\mathrm{ms}^{-1}$. The AVDS-car operation is in *velocity tracking mode* initially, and then *weak spacing control*.

Example 8.1: The AVDS-car is moving with the speed $20\mathrm{ms}^{-1}$. A car changes lanes in front of it reducing the safety distance by $10m$ but it is going faster than the AVDS-car by $6\mathrm{ms}^{-1}$. Thus the AVDS-car keeps original velocity (*velocity keeping mode*) until the spacing is satisfied. Then, it tracks the optimal velocity of $23\mathrm{ms}^{-1}$ (*velocity tracking mode*). Fig. 8.9 shows the control result.

Example 8.2: The AVDS-car is moving with speed 20ms^{-1}. A car with the speed 26ms^{-1} and acceleration -1ms^{-2} changes lane in front of it reducing the safety distance by 10m. Initially, the AVDS-car keeps its original velocity (*velocity keeping mode*) until that car's speed is less than the optimum velocity 23ms^{-1}. The AVDS-car uses *strong spacing control* to achieve the safety spacing. The speed profile of the cut-in car is shown in Fig. 8.10. Figure 8.11 shows the control result.

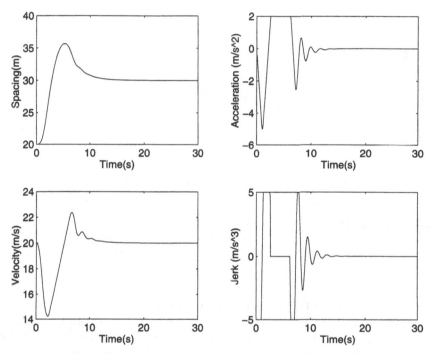

Fig. 8.13. The AVDS-car is moving at speed 20ms^{-1}. Suddenly, a car changes lanes in front of it at a distance which is 20m less than the required safety distance, with the speed 20ms^{-1}. The AVDS-car operation is in *weak spacing control*.

Example 8.3: The AVDS-car with speed 23ms^{-1} has no cars in its distance sensor for $t < 10$s. At $t = 10$s, it sees a car 60m ahead going at a speed 15ms^{-1}. The AVDS-car operation is in *velocity tracking mode* initially, and then *weak spacing control*. Figure 8.12 shows the control result.

Example 8.4: The AVDS-car is moving with the speed 20ms^{-1}. Suddenly, a car changes lanes in front of it at a distance which is 20m less than the required safety distance with the speed 20ms^{-1}. The AVDS-car operation is in *weak spacing control*. Figure 8.13 shows the control result.

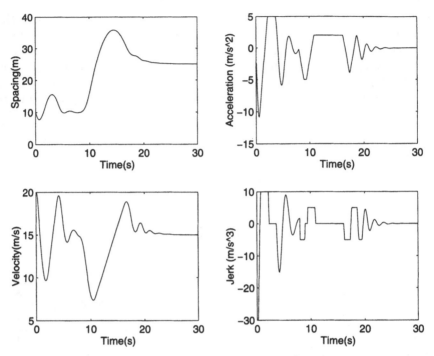

Fig. 8.14. The AVDS-car is moving with speed $20ms^{-1}$. A slow car changes lanes in front of it at a distance which is 10m less than the required safety distance with a lower speed of $15ms^{-1}$. The AVDS operation is in *emergency mode* initially, and then *weak spacing control*.

Example 8.5: The AVDS-car is moving with the speed $20ms^{-1}$. A slow car changes lanes in front of it at a distance which is 10m less than the required safety distance with the velocity going at a lower speed of $15ms^{-1}$. The AVDS operation is in *emergency mode* initially, and then *weak spacing control*. Figure 8.14 shows the control result.

Example 8.6: The AVDS-car has the closest position to a WANT-TO-CHANGE-LANE car in an adjacent lane, that is -8m. Thus, it determines to use a *creating mode* to create a space for that car to merge. Note that the initial spacing for this mode is less than zero. At time $t = 8s$, the WANT-TO-CHANGE-LANE car moves from the adjacent lane into the lane which the AVDS-car is driving in. Thus, the AVDS-car changes the driving mode to *weak spacing control*. Figure 8.15 shows the control result.

From these figures, it is observed that the AVDS system can handle various traffic scenarios, and achieve satisfactory control performance: the amplitudes of both jerk and acceleration are limited by the constraints, and tracking trajectories are very smooth.

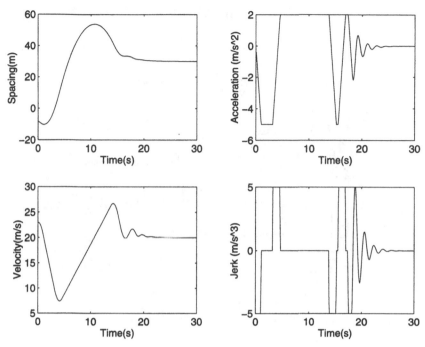

Fig. 8.15. The AVDS-car has the closest position to a WANT-TO-CHANGE-LANE car in an adjacent lane, that is -8m. Thus, it determines to use a *creating mode* to create a space for that car to merge. At time $t = 8s$, the WANT-TO-CHANGE-LANE car moves from the adjacent lane into the lane which the AVDS-car is driving in. Thus, the AVDS-car changes the driving mode to *weak spacing control*.

Remark 8.1

The computation time for the optimisation algorithm is about 0.01s on a Sun Workstation. It should be noted that the length of the prediction horizon plays a key role in the optimisation problem. When taking the AVDS here to a working test site, the prediction horizon should be considered as an off-line tuning parameter. For a larger value of P, the response of the output is slower and further decrease greater the overshoot. This brings passenger comfort. However, increasing the value of P results in an extension of the computation time. Thus, it is important to consider the cost of the computational time required to solve the optimisation due to hardware limitations. Also, the computation time should be less than the sampling instant. Therefore, in a practical design, one has to make a tradeoff between the cost of the computation time and the length of the control horizon.

8.3 Application to Injection Moulding Control

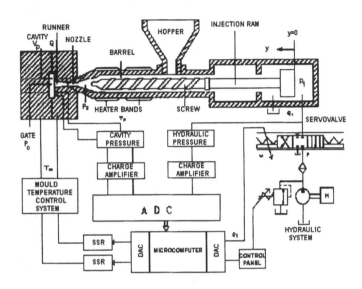

Fig. 8.16. Control injection moulding system

Injection moulding is one of the major processes used in the plastics manufacturing industry. Control of the injection moulding process involves three phases: filling, packing and holding, and cooling. The schematic of a typical injection moulding machine is shown in Fig. 8.16. Raw material (resin), at almost constant temperature in the hopper, is fed into the barrel as the screw revolves. The solid resin is heat melted through the iron layer from band heaters and by the frictional heat generated on the inside of the barrel. The resin moves gradually to the left in Fig. 8.16, because of the screw revolving. The molten resin is mixed and heated to a proper temperature uniformly. When a sufficient amount of molten resin accumulates in front of the screw, the ram moves quickly along the axis of the barrel, injecting the molten resin into the mould through the nozzle. As the resin begins to cool, a holding pressure on the screw is maintained to force more resin into the mould and 'pack ' the part into a compact unit. After the resin at the gateway to the mould solidifies, the holding pressure is removed and the part remains in the mould to cool. When the part is sufficiently cool, the mould opens and ejects the solid plastic part.

8.3.1 The Control Problem

For precise and consistent part production, it is very important to be able to accurately control the key injection moulding machine variables during each phase.

The problem is to design an advanced control algorithm to optimally track the position trajectory in the face of practical disturbances. With the trend towards production of increasingly precise miniature parts, conventional controllers such as fixed gain PID employed presently by the industry cannot achieve the tight control requirements. The main reason is that they do not utilise the known information about the future reference inputs, and therefore have sub-optimal performance. As a consequence the response is often characterised by unnecessarily large deviations from the reference profile. There have been many reports describing the application of advanced control techniques to this control problem. Pandelidis and Agrawal (1988) presented an optimal control of ram velocity assuming that the full states are available. Since the measurements from the available sensors often are corrupted by noise, Agrawal and Pandelidis (1988) further presented an application using the linear quadratic Gaussian (LQG) technique to estimate the process states. Zhang et al. (1996) suggested an adaptive control method for the ram velocity tracking problem. However, no detailed design procedure and theoretical analysis were provided in their paper. Based on a non-linear model, Tan et al. (2001) developed an adaptive sliding mode control for injection moulding machine.

In this section, PID control is designed using a GPC approach as in Chapter 5. It is simple to commission, accepting directly optimal performance weights, or alternatively, classical specifications from users which can be in turn converted into an initial performance index.

8.3.2 Process Model

As stated in the introduction, for this injection moulding machine, a simple second-order model is assumed as the control design basis:

$$P(s) = \frac{K_p}{(T_p s + 1)s},\qquad(8.52)$$

with K_p the static gain and T_p the time constant of the system.

From relay tuning the approximate plant model is

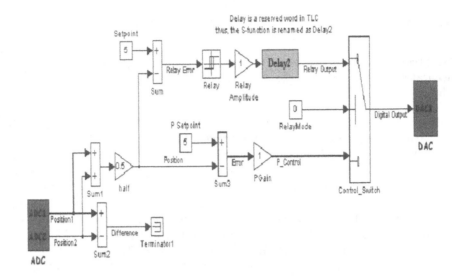

Fig. 8.17. SIMULINK for identifying model using relay tuning with delay

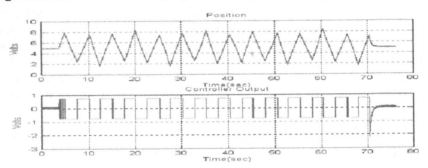

Fig. 8.18. Raw data for $L = 1.25 u_m = 0.75$

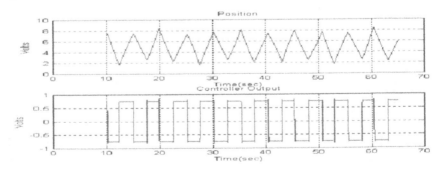

Fig. 8.19. Data range filtered and selected for identification $L = 1.25s$

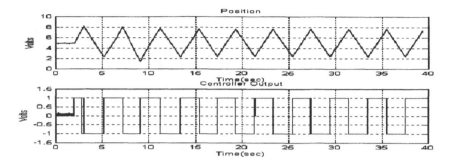

Fig. 8.20. Raw data for $L = 1, u_m = 1$

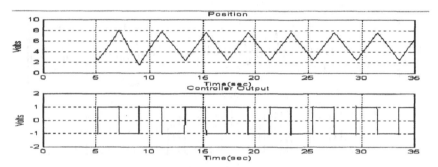

Fig. 8.21. Data range filtered and selected for identification $L = 1\text{s}$

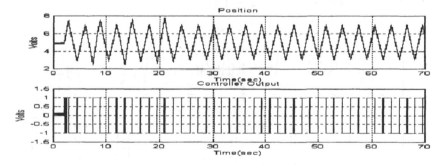

Fig. 8.22. Raw data for $L = 0.75, u_m = 1.25$

$$P(s) = \frac{3.492}{(0.01s + 1)s}. \tag{8.53}$$

The relay identification method for the injection moulding is considered for estimating the values of K_p and T_p and the scheme is shown in Fig. 8.17. The approach used is conventional relay tuning augmented with a pure time lag. For a second-order model given by (8.52), with the relay experiment conducted on the position loop, the parameters may be obtained as

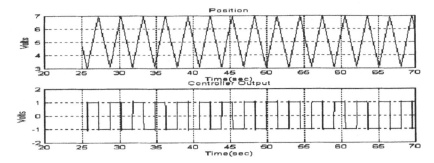

Fig. 8.23. Data range filtered and selected for identification $L = 0.75$s

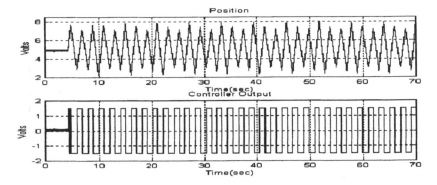

Fig. 8.24. Raw data for $L = 0.5, u_m = 1.5$

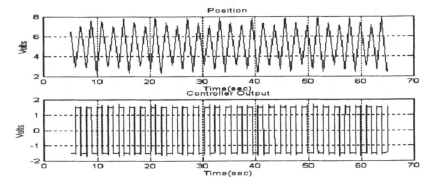

Fig. 8.25. Data range filtered and selected for identification $L = 0.5$s

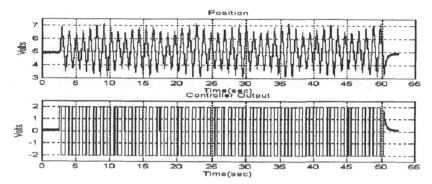

Fig. 8.26. Raw data for $L = 0.25, u_m = 2$

$$K_p = \frac{\omega^* \sqrt{1 + T_p^2 \omega*^2}}{K^*}, \tag{8.54}$$

$$T_p = \frac{cot(\omega^* L)}{\omega^*}. \tag{8.55}$$

If L is the additonal time delay introduced, the resultant phase angle shift

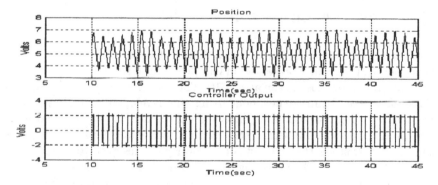

Fig. 8.27. Data range filtered and selected for identification $L = 0.25s$

of the negative inverse describing function can be shown to be $\omega^* L$, where ω^* is the frequency of the point of intersection between the inverse describing function of the relay-delay element and the Nyquist curve of the system. K^* is defined as $K^* = \frac{4u_m}{\pi y_m}$, where u_m is the amplitude of the relay and y_m is the oscillation amplitude.

Several sets of data were obtained for different values of delay and relay amplitude. The sampling instant used was 1ms for all cases. From the raw data collected, a portion was selected and low pass filtered with a cutoff frequency of 50 Hz before it was used for relay identification. This was done to remove any transient effect and measurement noise. The raw data and processed data are shown in Figs 8.18 and 8.27. The identification results are summarised in Table 8.1. Figures 8.21 and Fig. 8.23, with delay timing set to 1s and 0.75s, respectively, show that the limit cycle oscillations were uniform. Hence the K_p obtained from the two experiments is almost identical. For the rest, irregular oscillations repeat periodically. This is most prominent for a delay of 0.25s, as illustrated in Fig. 8.27. It is obvious that there were at least two limit cycles with different frequencies interfering with each other.

Table 8.1. Summary of relay tuning with delay

De(s)	Re(V)	T_u(s)	y_m	K^*	ω^*	T_p	K_p
1.25	0.75	5.0368	2.757	0.3463	1.2474	0.009207	3.6022
1	1	4.0515	2.584	0.4927	1.5508	0.01290	3.1479
0.75	1.25	3.0289	1.9287	0.6601	2.0744	0.007242	3.1426
0.5	1.5	2.0397	2.341	0.8158	3.080	0.009933	3.7776
0.25	2	1.0435	1.5989	1.5926	6.022	0.01088	3.789

De: Delay, Re: Relay, T_u: Periodic time of the output oscillation.

8.3.3 Controller Design and Experimential Results

Using the transfer function model (8.53), the discrete-time model can be obtained using the results presented in Chapter 5 with sampling instant 1ms. The prediction horizon is chosen as $p = 10$. The specifications $\omega_o = 15$ and $\zeta = 3$ are assumed. A real-time experiment using the developed control system is applied to an injection moulding machine comprosing an electro-hydraulic servo system built with industrial standard components. The servo system is essentially a test stand that consists of a power pack designed to power a low friction servo cylinder. The power pack has a pump output rated at $6cc^3 rev^{-1}$ and operates at a pressure of 40bar. It has a capacity of 60l of mineral oil. A 5.5kw electric motor is used as a prime mover to drive the hydraulic pump, which delivers flow to charge up a 1.4l accumulator. When the accumulator is charged, flow from the pump is unloaded back to the tank to prevent heat loss. During cylinder actuation, flow from the accumulator is directed to the servo cylinder via a servovalve mounted at the cylinder body. The position of the cylinder is detected by a potentiometer built within the cylinder. The control objective is to control the position of the cylinder using the predictive PID control. The schematic diagram of the test stand is shown in Fig. 8.28. A photograph of the experimental setup is shown in Fig. 8.29. The control algorithm is implemented using a National Instrument's PCI-MIO-16-E4 data acquisition card coupled with MATLAB's *Real-Time Windows Target*. The control performance is depicted in the top part of Fig. 8.30. For comparison, a PI controller tuned using the symmetrical optimum method (SO) (Åström and Hägglund, 1995) is shown in the lower part of Fig. 8.30.

The diagram clearly shows that position response is quick and smooth. The reason for this is that the predictive controller can take early action for any coming output change. Despite the fact that the predictive controller is based on a simple model of the plant, its performance is comparable to that of the PI controller.

8.4 Application to a DC Motor

Application of the adaptive predictive controller for the velocity control of a DC motor is discussed in this section. It is shown that the controller can easily be tuned by using the methods described in Chapter 7. An *MS15* DC Motor Control Module from L.J. Electronics is used as the test system. The experimental setup in the laboratory is shown in Fig. 8.31. A schematic of the DC motor module used is shown in Fig. 8.32. The module consists of a DC motor that is capable of being driven at speeds up to 2500rpm in either direction. A second DC motor driven directly by the first provides an analogue voltage feedback proportional to the speed and direction of rotation. The controller was implemented with MATLAB's SIMULINK with *Real Time Windows Target* on a National Instruments data acquisition card *PCI-MIO-16E-4*. The sampling period was 0.001s.

The controller, shown in Fig. 8.33, essentially consists of two major block. One block performs recursive identification using the differential input and output. A second block calculates the controller output based set-point, system's output, user set control horizon and the parameter vector identified from the first block. There are two variables to initialise for the recursive algorithm. They are the value of ϱ, and the initial estimation vector, $\phi(k)$. Choose $\varrho = 0.96$. For practical real-time implementation, the initial estimation vector $\phi(k)$ is set to be an all zero vector.

The response of the system undergoing step changes in set-point in intervals of 10 seconds is shown in Fig. 8.34 and Fig. 8.35 with input weighting factor, r set to 1 and 8, respectively, when the prediction horizon $p = 6$ is fixed. Disturbance was introduced by lifting the Eddy Current Brake manually at around 25 seconds. The two figures show that smaller r results in faster response, larger overshoot and better disturbance rejection.

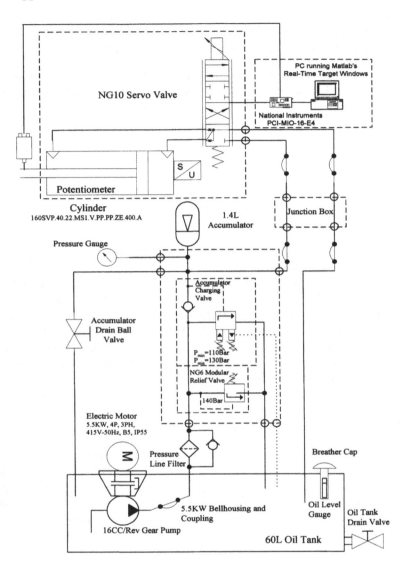

Fig. 8.28. Schematic diagram of the electro-hydraulic servo system

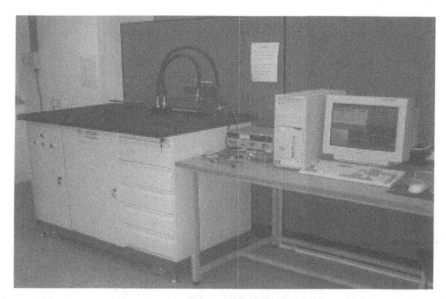

Fig. 8.29. Schematic diagram of the electro-hydraulic servo system

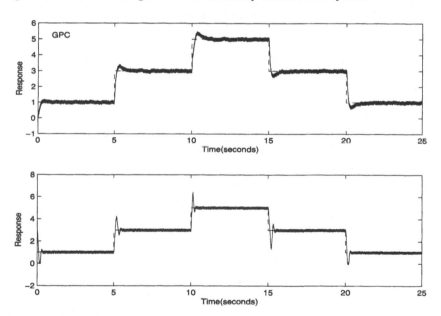

Fig. 8.30. Real-time response

Fig. 8.31. Experimental setup in the laboratory

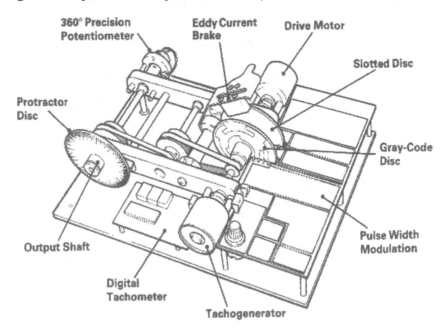

Fig. 8.32. A schematic of the DC motor module used in the real-time experiment

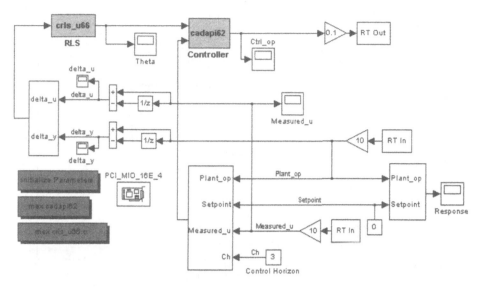

Fig. 8.33. SIMULINK model used in the real-time experiment

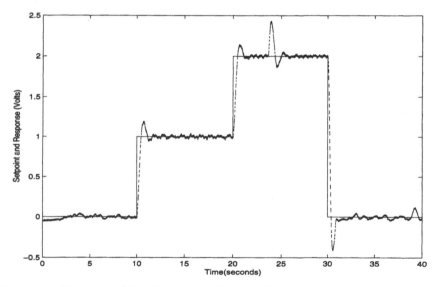

Fig. 8.34. Response of the DC motor with r set to 1

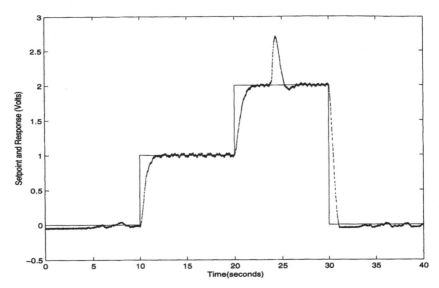

Fig. 8.35. Response of the DC motor with r set to 8

REFERENCES

Agrawal,A.R. and I.O.Pandelidis, Observers for optimal anticipatory control of ram velocity in injection molding, *Polymer Engineering and Science*, vol.28, pp.157–164, 1988.

Agrawal,A.R., I.O. Pandelidis and M.Pecht, Injection-molding process control- A review, *Polymer Engineering and Science*, vol.27, pp.1345–1357, 1987.

Albertos,P. and R.Ortega, On generalised predictive control: Two alternative formulations, *Automatica*, vol.25, no.5, pp.753–755, 1989.

Alevisakis,G. and D.E.Seborg, Control of multivariable systems containing time delays using a multivariable Smith predictor, *Chemical Engineering Science*, vol.29, pp.373–380, 1974.

Allidina,A.Y. and F.M.Hughes, Generalised self-tuning controllers with pole assignment, *Proceedings of IEE, Part D*, vol.127, pp.13–18, 1980.

Alvarez,T. and C.Prada, Handling infeasibility in predictive control, *Computers and Chemical Engineering*, vol.21, pp.577–582, 1997.

Amann,N., D.H.Owens and E.Rogers, Iterative learning control for discrete-time system with exponential rate of convergence, *IEE Proceedings, Part D*, vol.143, no.2, pp.217–224, 1996.

Amann,N., D.H.Owens and E.Rogers, Predictive optimal iterative learning control, *International Journal of Control*, vol.69, pp.203–226, 1998.

Arimoto,S., S.Kawamura and F.Miyazaki, Bettering operating operation of robots by learning, *Journal of Robotic Systems.*, vol.1, pp.123–140, 1984.

Åström,K.J., *Introduction to Stochastic Control Theory*, Academic Press, New York, 1970.

Åström,K.J. and T.Hägglund, *PID Controllers: Theory, Design, and Tuning*, 2nd Edition, Instrument Society of America, 1995.

Åström,K.J., T.Hägglund, C.C.Hang and W.K.Ho, Automatic tuning and adaptation for PID controllers-A survey, *Control Engineering Practice*, vol.1, pp.699–714, 1993.

Bellman,R. and K.L.Cooke, *Differential-Difference Equation*, Academic Press, 1963.

Bender,J., An overview of systems studies of automated highway systems, *IEEE Transactions on Vehicular Technology*, vol.40, pp.82–99, 1991.

Bien,Z. and K.M.Huh, High-order iterative control algorithm, *IEE Proceedings, Part-D, Control Theory and Applications*, vol.13, no.3, pp.105–112, 1989.

Burl,J.B., *Linear Optimal Control: H_2 and H_∞ Methods*, Addison-Wesley Longman, Inc.,1999.

Camacho,E.F. and C.Bordons, *Model Predictive Control*, Springer-Verlag London, 1999.

Casavola,A., M.Giannelli and E.Mosca, Min-Max predictive control strategies for input-saturated polytopic uncertain systems, *Automatica*, vol.36, pp.125–133, 2000.

Chen,C.T., *Linear System Theory and Design* (3rd edition), Oxford University Press, Inc., New York, 1999.

Chen,F.C. and H. K. Khalil, Adaptive control of a class of nonliear discrete-time system using neural networks, *IEEE Transaction on Automatic Control*, vol.AC-40, no.5, pp.791–801, 1995.

Chen,Y.Q. and C.Wen, *Iterative Learning Control: Convergence, Robustness, and Applications*, Springer-Verlag, 1999.

Chidambaram,M., Design of PI controllers for integrator/dead-time processes, *J. of Inds. Chem.*, vol. 22, pp.37–39, 1994.

Clarke,D.W., C.Mohtadi and P.S.Tuffs, Generalized predictive control-Part 1, Basic algorithm, *Automatica*, vol.23, pp.137–148, 1987.

Clarke,D.W. and C.Mohtadi, Properties of generalised predictive control, *Automatica*, vol.25, no.6, pp.859-875, 1989.

Clarke,D.W., E.Mosca and R.Scattolini, Robustness of an adaptive predictive controller, *IEEE Transactions on Automatic Control*, vol.39, no.5, pp.1052-1056, 1994.

Clarke,D.W. and R.Scattolini, Robustness of an adaptive predictive controller, *Proceedings of the 30th Conference on Decision & Control*, pp.979–984, 1991.

Clarke,D.W. and R.Scattolini, Constrained receding-horizon predictive control, *IEE Proceedings Part D.*, vol.138, no.4, pp.347–354, 1991.

Cutler,C.R. and B.L.Ramaker, Dynamic matrix control - A computer control algorithm, *Proceedings JACC*, San Francisco, USA, paper WP5-B, 1980.

De Keyser,R. and A.Van Cauwenberghe, A self-tuning multistep predictor application, *Automatica*, vol.17, pp.167–174, 1981.

De Keyser,R., Ph. Van de Velde and F. Dumortier, A comparative study of self-adaptive long-range predictive control methods, *Automatica*, vol.24, pp.149–163, 1988.

de Paor,A. M. and M. O'Malley, Controllers of Ziegler-Nichols type for unstable process with time delay, *International Journal of Control*, vol. 49, no.4, pp.1273-1284, 1989.

Draeger,A., S.Engell and H.Ranke, Model predictive control using neural networks, *IEEE Control Systems*, vol.15, no.5, pp.61–66, 1995.

Elshafei,A., G. Dumont and A. Elnaggar, Stability and convergence analyses of an adaptive GPC based on state-space modelling, *International*

Journal of Control, vol.61, no.1, pp.193–210, 1995.

Eskafi,F., D.Khorramabadi and P.Varaiya, Smartpath: An automated high-way system simulator, *TECH.REP.PATH Techical Note 94-3*, 1994.

Fisher,M., O.Nelles and R.Isermann, Predictive control based on local linear fuzzy models, *International Journal of Systems Science*, vol.29, no.7, pp.679–697, 1998.

Fujimoto,Y. and A.Kawamura, Robust servo-system based on two-degree-of-freedom control with sliding mode, *IEEE Transactions on Industrial Electronics*, vol.42, no.3, pp.272–280, 1995.

Furukawa,F. and E. Shimemura, Predictive control for systems with delay, *International Journal of Control*, vol.37, pp.399–412, 1983.

Gambier,A. and H.Unbehauen, Multivariable generalized state-space receding horizon control in a real-time environment, *Automatica*, vol.35, pp.1787–1797, 1991.

Garcia,C.E. and M.Morari, Internal model control: 1.An unifying review and some new results, *Industrial & Engineering Chemistry Process Design and Development*, vol.21, pp.308–323, 1982.

Garcia,C.E. and A.M.Morshedi, Solution of the dynamic matrix control problem via quadratic programming, *Proceedings of the Conference of the Canadian Industrial Computing Society*, Ottawa, Canada, pp.1–13, 1984.

Geromel,J.C., P.L.D.Peres and J.Bernoussou, On a convex parameter space method for linear control design of uncertain systems, *SIAM Journal of Control and Optimization*, vol.29, no.2, pp.381–402, 1991.

Goh, C.J., Model reference control of nonlinear systems via implicit function emulation, *International Journal of Control*, vol.60, no.1, pp.91–115, 1994.

Goodwin,G. and K.Sin, *Adaptive Filtering, Prediction and Control*, Prentice Hall, Englewood Cliffs, NJ, 1984.

Hägglund,T. A predictive PI controller for processes with long dead times, *IEEE Control Systems Magazine*, vol. 12, no.1, pp.57–60, 1992.

Hägglund,T. and K.J. Åström, Industrial adaptive controllers based on frequency response techniques, *Automatica*, vol.27, pp.599–609, 1991.

Haykin,S., *Adaptive Filter Theory*, Third Edition, Prentice Hall Inc., 1996.

Hernandez,E. and Y.Arkun, Neural network modeling and an extended DMC algorithm to control nonlinear systems, *Prococeedings of 1990 American Control Conference*, pp.2454–2459, 1990.

Ho, W. K. and W.Xu, PID tuning for unstable processes based on gain and phase specifications, *IEE Proceedings, Part D*, vol.145, no.5, pp.392–396, 1998.

Horn,R. A. and C.R.Johnson, C. R., *Matrix Analysis*, Cambridge University Press, New York, 1985.

Hou, Z. and W. Huang, The model-free learning adaptive control of a class of SISO nonlinear systems, *Proceedings of the American Control Conference*, Albuquerque, New Mexico, pp.343–344, 1997.

Hrovat,D., Optimal suspension performance for 2-D vehicle models, *Journal of Sound and Vibration*, vol.146, no.1, pp.93–110, 1991.

Huang,H. P. and C.C.Chen, Control system synthesis for open-loop unstable process having time delay, *IEE Proceedings, Part D*, vol. 144, no.4, pp.334–346, 1997.

Huang,S.N. and W.Ren, Safety, comfort, and optimal tracking control in AHS applications, *IEEE Control Systems*, vol.18, pp.50–64, 1998.

Huang,S.N., K.K.Tan and T.H.Lee, Generalized predictive observer-controller for injection molding with input delay and non-measurable noise, *International Polymer Processing*, vol. XIV, pp.399–408, 1999a.

Huang,S.N., K.K.Tan,and T.H.Lee, Predictive control of ram velocity in injection molding, *Polymer-Plastics Technology and Engineering*, vol.38, pp.285–304, 1999b.

Huang,S.N., K.K.Tan and T.H.Lee, Adaptive GPC control of melt temperature in injection moulding, *ISA Transactions*, vol.38, pp.361–373, 1999c.

Huang,S.N., K.K.Tan and T.H. Lee, A combined PID/adaptive controller for a class of nonlinear systems,*Automatica*, vol.37,no.4, pp.611–618,2001.

Ichikawa,K., Frequency-domain pole assignment and exact model-matching for delay systems, *International Journal of Control*, vol.41, pp.1015–1024, 1985.

Ioannou,P. and C.Chien, Autonomous intelligent cruise control, *IEEE Transactions on Vehicular Technology*, vol.42, pp.657–672, 1993.

Ito, K., T. Fujishiro, K. Kanai and Y. Ochi, Stability analysis of aut014tic lateral motion controlled vehicle with four wheel steering system, *Proceedings of the 1990 American Control Conference*, San Diego, May, pp.801–808, 1990.

Jacquot,R.G., *Modern Digital Control Systems*, Marcel Dekker, Inc. New York, 1981.

Jordan,M.I. and D. E. Rumelhart, Forward models: supervised learning with a distal teacher, *Cognitive Science*, vol.16, pp.307–354, 1992.

Kalman,R.E., A new approach to linear filtering and prediction problems, *Transactions of ASME, Journal of Basic Engineering*, vol.82, pp.35–45, 1960.

Kaya,I.and D.P.Atherton, A PI-PD controller design for integrating processes, *Proceedings of the American Control Conference*, San Diego, California, 1999.

Kouvaritakis,B., J.A.Rossiter and A.O.T.Chang, Stable generalized predictive control: An algorithm with guaranteed stability, *IEE Proceedings, Part D*, vol.139, no.4, pp.349–362, 1992.

Kroyszig,E., *Advanced Engineering Mathematics*, Seventh Edition, John Wiley & Sons,Inc.,New York, 1993.

Kwon,W.H, Y.Lee and S.Noh, Partition of GPC into a state observer and a state feedback controller, *Proceedings of 1992 American Control Conference*, Illinois, pp.2032–2041, 1992.

Laughlin,D.L., D. Rivera and M. Morari, Smith predictor design for robust performance,*International Journal of Control*, vol.46, no.2, pp.477–504, 1987.

Lee, J.H., K.S.Lee and W.C. Kim, Model-based iterative learning control with a quadratic criterion for time-varying linear systems, *Automatica*, vol.36, pp.641–657, 2000.

Lee, J.H., M.S. Gelormino and M. Morari, Model predictive control of multi-rate sampled-data systems: A state-space approach, *International Journal of Control*, vol.55, pp.153–191, 1992.

Lee, J.H., M. Morari and C.E. Garcia, State-space interpretation of model predictive control, *Automatica*, vol.30, no.4, pp.707–717, 1994.

Lee, J.H. and Z.H. Yu, Robust tuning parameters for model predictive control, *Computers & Chemical Engineering*, vol.18, no.1, pp.15–37, 1994.

Lee, J.H. and Z. Yu, Worst-case formulations of model predictive control for systems with bounded parameters, *Automatica*, vol.33, no.5, pp.763–781, 1997.

Lee, K.S. and J.H.Lee, Model-based predictive control combined with iterative learning for batch processes, Edited by Z.Bien and J.X.Xu, Kluwer Academic, pp.314–334, 1999.

Li, S., K.Lim and D.Fisher, A state-space formulation for model predictive control, *AICHE Journal*, vol.35, pp.241–249, 1989.

Lu, Y. and Y.Arkun, Quasi-Min-Max MPC algorithms for LPV systems, *Automatica*, vol.36, pp.527–540, 2000.

Luyben,W.L.,*Process Modeling, Simulation and Control for Chemical Engineers*, (2nd Edition) McGraw-Hill Publishing Company, 1990.

Luyben,W.L., Tuning PI controllers for processes with both inverse response and deadtime, *Industrial and Engineering Chemistry Research*, vol.39, no.4, pp.973–976, 2000.

Manitius,A. and A.W.Olbrot, Finite spectrum assignment problem for system with delay, *IEEE Transactions on Automatic Control*, vol.24, pp.541–553, 1979.

Marshall,J.E., H.Gorecki, K.Walton and A.Korytowski, *Time-Delay Systems: Stability and Performance Criteria with Applications*, Ellis Horwood, 1992.

Martin Sanchez,J.M. and J.Rodellar, *Adaptive Predictive Control: From the Concepts to Plant Optimization*, Prentice Hall, 1996.

Mayne,D. and H.Michalska, Receding horizon control of nonlinear systems, *IEEE Transactions on Automatic Control*, vol.35, pp.814–824, 1990.

Mayne,D. and H.Michalska, Robust receding horizon control of constrained nonlinear systems, *IEEE Transactions on Automatic Control*, vol.38, pp.1623–1633, 1993.

Moore,K.L., *Iterative Learning Control for Deterministic Systems*, Advances in Industrial Control Series, London, UK: Springer-Verlag, 1992.

Morari,M. and E.Zafiriou, *Robust Process Control*, Prentice-Hall Inc., 1989.

Mosca,E. and J.Zhang, Stable redesign of predictive control, *Automatica*, vol.28, no.6, pp.1229–1233, 1992.

Murrill,P.W., *Application Concepts of Process Control*, Instrument Society of America Press, Research Triangle Park, NC, 1988.

Narendra,K.S. and K. Parthasarathy, Identification and control of dynamical systems using neural networks, *IEEE Transactions on Neural Networks*, vol.1, pp.4–27, 1990.

Nasar,S.A. and I.Boldea, *Linear Electric Motors:Theory, Design and Practical Applications*, Prentice-Hall, Inc., 1987.

Niehaus,A. and R.E.Stengel, Probability-based decision making for automated highway driving, *IEEE Transactions on Vehicular Technology*, vol.43, pp.626–634, 1994.

Palmor,Z.J. and R. Shinnar, Design of advanced process controllers, *AIChe Journal*, vol.27, no.5, pp.793–805, 1981.

Pandelidis,I.O. and A.R.Agrawal, Optimal anticipatory control of ram velocity in injection molding, *Polymer Engineering and Science*, vol.28, pp.147–156, 1988.

Payne,D.Q, Stability result with application to adaptive control, *International Journal of Control*, vol.46, no.1, pp.249–261, 1987.

Peterka,V., Predictor-based self-tuning control. *Automatica*, vol.20, no.1, pp.39–50, 1984.

Peterson,T., E.Hernandez, Y.Arkun and F.J.Schork, Nonlinear predictive control of a semi batch polymerisation reactor by an extended DMC, *Proceedings of 1989 American Control Conference*, pp.1534–1539, 1989.

Piche,S., B.Sayyar-Rodsari, D.Johnson and M.Gerules, Nonlinear model predictive control using neural networks, *IEEE Control Systems*, vol.20, pp.53–62, 2000.

Rafizadeh,M., W. I. Patterson and M. R. Kamal, Physically-based model of thermoplastics injection molding for control applications, *International Polymer Processing*, vol.11, no.4, pp.352–361, 1996.

Rawlings,J.B. and K.R.Muske, The stability of constrained receding horizon control, *IEEE Transactions on Automatic Control*, vol.38, pp.1512–1516, 1993.

Richalet,J., A.Rault, J.L.Testud and J.Papon, Model predictive heuristic control: applications to industrial processes, *Automatica*, vol.14, no.5, pp.413–428,1978.

Ricker,N.L., Model predictive control with state estimation, *Industrial and Engineering Chemical Research*, vol.29, pp.371–382, 1990.

Rigopoulos,A., Y.Arkun and F.Kayihan, Model predictive control of CD profiles in sheet forming processes using full profile disturbance models identified by adaptive PCA, *Proceeding of the American Control Conference*, Albuquerque, New Mexico, June, pp.1468–1472, 1997.

Rodellar,J., Optimal design of the driver block in the adaptive predictive control system (in Spanish), Doctoral dissertation, University of Barcelona, Spain, 1982.

Rossiter,J.A. and B.Kouvaritakis, Constrained stable generalised predictive control, *IEE Proceedings, Part D.*, vol.140, pp.243–254, 1993.

Rotstein,G.E. and D.R.Lewin, Simple PI and PID tuning for open-loop unstable systems,*Ind. Eng. Chem. Res.*, vol.30, pp.1864–1869, 1991.

Saab,S.S., A discrete learning control algorithm for a class of linear time-invariant systems, *IEEE Transactions on Automatic Control*, vol.40, no.6, pp.1138–1141, 1995.

Scattolini,R. and S.Bittanti, On the choice of the horizon in long-range predictive control-some simple criteria, *Automatica*, vol.26, pp.915–917, 1990.

Scokaert,P.O.M., Infinite horizon generalised predictive control, *International Journal of Control*, vol.66, no.1, pp.161–175, 1997.

Shiro,U., K.Takuya, S.Takayuki and H.Hironobu, High performance positioing system with linear dc motor under self-tuning fuzzy control, *The Second International Symposium on Linear Drives for Industry Applications*, Tokyo, Japan, pp.295–298, 1995.

Smith,C.A. and A.B.Corripio, *Principles and Practice of Automatic Process Control*, John Wiley & Sons, 1985.

Smith,O.J.M., A controller to overcome dead-time, *ISA Transactions*, vol.6, no.2, pp.28–33, 1959.

Soeterboek,R., *Predictive Control:A Unified Approach*, Prentice Hall International (UK) Limited, 1992.

Stahl,H. and P.Hippe, Design of pole placing controllers for stable and unstable systems with pure time delay, *International Journal of Control*, vol. 45, no.6, pp.2173-2182, 1987.

Sung,S.W. and J.O.J.Lee, S.Yu and I.Lee, Automatic tuning of PID controller using second order plus time delay model, *Journal of Chemical Engineering*, vol.29, pp.990–999, 1996.

Tan, K.K., H.F. Dou, Y.Q. Chen and T.H. Lee, High precision linear motor control via relay tuning and iterative learning based on zero-phase filtering, *IEEE Transactions on Control Systems Technology*, vol.9, no.2, pp.244–253, 2001.

Tan, K.K., S.N.Huang and T.H.Lee, Development of a GPC-based PID controller for unstable systems with deadtime, *ISA Transaction*, vol.39, no.1, pp.57–70, 2000.

Tan, K.K., S.N.Huang and T.H.Lee, Learning-enhanced generalized predictive control for systems with periodic disturbances, *Systems and Control Letters* (Under Review)

Tan, K.K., S.N.Huang, T.H.Lee and F.M.Leu, Adaptive-predictive PI control of a class of SISO system, *Proceedings of the American Control Conference*, pp.3848–3852, 1999.

Tan, K.K. and T.H.Lee, Automatic tuning of PID cascade controllers for servo systems, *Intelligent Automation and Soft Computing*, vol.4, no.4, pp.325–340, 1998.

Tan, K.K., T.H.Lee and R.Ferdous, New approach for design and automatic tuning of the Smith predictor controller, *Industrial and Engineering Chemistry Research*, vol.38, no.9, pp.3438–3445, 1999.

Tan, K.K., T.H.Lee and S.N.Huang, Adaptive control of ram velocity in injection molding, *IEEE Transactions on Control Systems Technology*, vol.9, no.4, pp.663–671, 2001.

Tan, K.K., T.H.Lee and S.N.Huang, Predictive iterative learning control, *in Proceedings of The 4th Asian Control conference*, Singapore, 2002.

Tan, K.K., T.H.Lee, S.N.Huang and X.Jiang, Optimal PI control for time-delay systems based on a GPC approach, Internal Report #EE-CNTL-98-01, National University of Singapore, 1998.

Tan, K.K., T.H.Lee, S.N.Huang and F.M.Leu, Adaptive-predictive control of a class of SISO nonlinear systems, *Dynamics and Control* (Accepted).

Tan, K.K., T.H.Lee, S.N.Huang and F.M.Leu, PID control design based on a GPC approach, *Automatica* (Under Review)

Tan, K.K., T.H.Lee, S.N.Huang and Q.G.Wang, A novel predictive and self-tuning using PI control apparatus for expanded process application, Singapore Patent, No.9804845–7, 1999.

Tan, K.K., T. H. Lee, S. Y. Lim and H. F. Dou, Learning enhanced motion control of permanent linear motor, *Proceedings of IFAC International Workshop on Motion Control*, Grenoble, France, pp.397–402, 1998.

Tan, K.K., T.H.Lee and F.M.Leu, Predictive PI versus Smith control for dead-time compensator, *ISA Transactions*, vol.40, no.1, pp.17–29, 2001.

Varaiya,P., Smart cars on smart roads: problems of control, *IEEE Transactions on Automatic Control*, vol.38, pp.195–207, 1993.

Venkatashankar,V. and M.Chidambaram, Design of P and PI controllers for unstable first-order plus dead time delay systems, *International Journal of Control*, vol. 60, no.1, pp.137–144, 1994.

Walzer,P. and W.Zimdahl, European concepts for vehicle safety, communication and guidance, *Proceedings of International Congress on Transportation Electronics*, pp.91–95, 1988.

Wang,K.K., Injection Molding Modelling, Cornell University, Progress Report, 1984.

Wang,L. and W.R.Cluett, Tuning PID controllers for integrating processes, *IEE Proceedins, Part D*, vol. 144, no.5, pp.385–392, 1999.

Warwick,K.W. and D.Rees, *Instrumential Digital Control System,* Peter Peregrinus, London, 1986.

Wei, J.H., C. C. Chang and C. P. Chiu, A nonlinear dynamic model of a servo-pump controlled injection molding machine, *Polymer Engineering and Science*, vol.34, no.11, pp.881–887, 1994.

Wolovich,W.A., *Linear Multivariable Systems*, Springer-Verlag:Berlin, 1974.

Yamanaka,K. and E.Shimemura, Effects of mismatched Smith controller on stability in systems with time delay, *Automatica*, vol.23, no.6, pp.787–791, 1987.

Ydstie,B.E., Extended horizon adaptive control, *Proceedings of the 9th IFAC World Congress*, Budapest, Hungary, 1984.

Yoon,T. and D.Clarke, Receding-horizon predictive control with exponential weighting, *International Journal of Systems Sciences*, vol.24, pp.1745–1757, 1993.

Yoon,T. and D.Clarke, Observer design in receding-horizon predictive control, *International Journal of Control*, vol.61, pp.171–191, 1995.

Zelinka,P., B.Rohal-Ilkiv and A.G.Kuznetsov, Experimental verification of stabilising predictive control, *Control Engineering Practice*, vol.7, pp.601-610, 1999.

Zheng,A. and M. Morari, Robust stability of constrained model predictive control, *Proceedings of American Control Conference*, pp.379–383, San

Francisco, CA, 1993.

Zhang,C.Y., J. Leonard and R.G. Speight, Adaptive controller performance used for ram velocity control during filling phase, *SPE Annual Technology Papers*, vol.42, pp.593–598, 1996.

INDEX

Acceleration Constraints, 224
Accelerations, 224
Actuators, 46
Adaptive Controller, 183
Adaptive Predictive Control, 181

Bandwidth Constraint, 15
Batch, 156

CARIMA, 74, 182, 183
Characteristic Equation, 27, 142
Characteristic Polynomial, 17
Closed-loop Poles, 20
Closed-loop System, 47, 124
Compensator, 9
Constraints, 11, 46, 70, 180, 224
Control Action, 40
Control Horizon, 5, 141, 223
Convergence, 160, 172
Cost Function, 39
Criterion Function, 37

Damping Ratio, 111, 143
DC Motor, 61, 208, 243
Dead Time, 9
Delay Systems, 17
Delay-free System, 17
Discrete-time Model, 89
Disturbance Learning, 53, 54
Disturbance Model, 61
Dynamic Linearisation, 68, 177
Dynamic Matrix Control, 207

Error Equation, 86
Estimator, 183
Extended Horizon Adaptive Cotrol, 182
Extended Prediction Self-adaptive
    Control, 182

Feedback Control, 12
Feedforward Estimate, 56
Feedforward Learning, 57, 60

Filter, 78
Finite Spectrum Assignment, 2, 16
First-order Model, 134
First-order Systems, 106
Free System, 82
Frequency Domain, 16
Future Outputs, 137, 223
Future States, 138

Generalised Predictive Control (GPC),
    4, 37, 53, 100, 101, 207

Hard Constraints, 46
Hardware Constraints, 225
High-order System, 150
Hurwitz Polynomials, 24

Identification, 156
Identification Algorithm, 182, 191
Injection Moulding, 88, 129, 165, 202,
    236
Input Constraints, 46
Input Saturation, 46, 48
Intelligent Autonomous Vehicle, 208
Internal Model Control, 11
Inverse, 48
Iterative Learning Control (ILC), 54,
    57, 153, 154

Kalman Filter, 76

Laplace Transform, 20, 24
Linear Inequality Constraints, 224
Linear Motor, 204
Linear Systems, 136
Load Disturbance, 25, 134
Long-range Predictive Control, 3
Luenberger Observer, 17
Lyapunov Equation, 87, 162
Lyapunov Function, 43, 194

Natural Frequency, 111, 143

Necessary and Sufficient Condition, 14
Neural Network, 65, 177
Nominal Plant, 13
Nominal Stability, 12, 85
Nominal System, 24, 86
Non-linear Model, 178
Non-linear Predictive Control, 62
Non-linear Predictive Iterative Learning
    Control, 175
Non-linear System, 184
Non-measurable Disturbances, 73
Non-minimum Phase System, 133
Nyquist, 33

Observer, 73, 76, 159
Optimal Control, 138
Optimal PI Controller, 101
Optimal PID Controller, 135
Optimisation, 84, 158, 223, 224
Output Constraints, 225

Parameter Estimation, 187
Parameter Identification, 182, 191
Partial Derivative, 184
Performance Criterion, 39, 41
Performance Index, 39
Periodic Disturbance, 58
Perturbed System, 42, 85, 125
PI Controller, 106
PID, 99, 100, 207
Pole Assignment, 18
Poles, 13
Practical Stability, 27
Predicted State, 18, 138
Prediction, 138
Prediction Equation, 76
Prediction Horizon, 4, 39, 40, 56, 75,
    137
Prediction Interval, 39, 53, 54, 75, 157,
    222
Prediction Model, 54, 74, 157, 222
Predictive Control, 1, 37, 221
Predictive Iterative Learning Control,
    153
Predictor, 156, 170

Quadratic Performance, 55, 75, 157,
    188

Quadratic Programming, 224

Receding Horizon, 138
Recursive Identification, 183
Recursive Least squares Algorithm,
    183, 188
Reference Trajectory, 39, 55, 75, 189,
    211
Regulation, 21
Repetitive Operation, 155
Riccati Equation, 77
Robust Performance, 15
Robust Stability, 11, 32
Robustness, 84, 124, 161, 163

Sampling Instant, 39, 54, 75, 221
Sampling Time, 138
Saturating Actuators, 46
Saturating Control, 49
Sensitivity Function, 15, 34
Set-point, 11, 25, 134
SISO, 12, 43, 73, 181
Smith-predictor Controller, 1, 9
Soft Constraints, 46
Stability, 42, 47, 80, 105, 113, 191
State Constraints, 46

Taylor Series, 68, 178
Time Delay, 11
Time Domain, 16, 17
Tracking, 21
Transfer Function, 11, 16, 27, 68, 135,
    242

Uncertainty, 161
Uncertainty Regions, 14
Uncertainty Set, 13
Unstable Poles, 33
Unstable System, 148

Violations, 46

Weighting Matrices, 141
Worst Case, 15

Ziegler–Nichols, 100